T0221550

The Joy of SET®

The Joy of SET

The Many Mathematical Dimensions of a Seemingly Simple Card Game

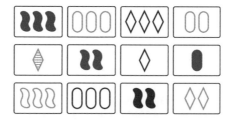

Liz McMahon, Gary Gordon,
Hannah Gordon, Rebecca Gordon

PRINCETON UNIVERSITY PRESS
PRINCETON AND OXFORD
NATIONAL MUSEUM OF MATHEMATICS
NEW YORK

Jacket image courtesy of Set Enterprises, Inc.

All Rights Reserved
Library of Congress Cataloging-in-Publication Data

Names: McMahon, Liz (Elizabeth Wagner)

Title: The joy of set : the many mathematical dimensions of a seemingly
simple card game / Liz McMahon [and three others].
Description: Princeton : Princeton University Press, [2017] | Includes
bibliographical references and index.
Identifiers: LCCN 2016016189 | ISBN 9780691166148 (hardcover : alk. paper)
Subjects: LCSH: Mathematical recreations. | Card games.
Classification: LCC QA95 .J875 2017 | DDC 793.74–dc23 LC record available
at https://lccn.loc.gov/2016016189

British Library Cataloging-in-Publication Data is available

This book has been composed in Minion Pro

Printed on acid-free paper. ∞

press.princeton.edu

Typeset by Nova Techset Pvt Ltd, Bangalore, India
Printed in the United States of America

1 3 5 7 9 10 8 6 4 2

CONTENTS

We love SET®. The game is easy to learn, and it can be played at a very high level by young children, while adults also find it fun and challenging. Indeed, as the authors of this book can attest, children can quickly surpass adults in speed and accuracy of identifying *SET*s (and stay ahead forever). The game is also popular in schools, ranging from enrichment at the elementary school level to college math clubs. At its root, the game is about pattern recognition and matching. That simple observation connects the game with mathematics, but as we'll see, the mathematics goes much deeper.

SET is one of the most popular games of the last 25 years. Since its release in 1991, it has been recognized repeatedly as an outstanding educational game (the Set Enterprises website lists at least 37 awards spanning more than 20 years).

The game has also attracted the interest of the mathematics community. A wide variety of material (covering a wide variety of levels) has been produced that is related to the game. This includes both research and pedagogical articles about using SET in the classroom, and projects related to specific courses in geometry, abstract algebra, linear algebra, and combinatorics. Many questions that arise naturally when playing the game are mathematical in nature; some of these have been addressed on the web in different forms, occasionally with conflicting claims.

Why Did We Write This Book?

We repeat, we love SET. We've been playing the game for years, both as a family and with friends. We've introduced the game to a wide variety of people, and all of us have spoken about some of the topics in this book in classes and at seminars and conferences. Audiences respond enthusiastically, and we believe the game has a universal appeal.

As teachers, we have also used SET as a fun way to introduce certain topics in high-school and college math classes, including upper-level college courses in geometry and combinatorics. People love games, and SET is a terrific way to engage students. When students begin to play SET, they start asking the "right" questions, learning more about mathematics and the game at the same time.

One of our goals is to help people understand that a mathematical approach to this game can have big payoffs. This is a two-way street: knowing some mathematics can enhance your understanding of the game, and playing SET can enhance your understanding of mathematics. Moreover, while there is material available on the web and in articles, this book is the first unified treatment of the mathematics connected to the game.

Another important goal we have is for people to realize that mathematics is everywhere, and if you look for it, you'll find that mathematics can help you understand the world. Mathematics, fundamentally, is about patterns, and patterns are all around us. Nearly everyone can do math, just as nearly everyone can read and write. We hope that, in actively reading this book, you will confidently consider yourself a "math person," regardless of your mathematical background.

The Game

The game of SET uses a special deck of cards. Each card has symbols characterized by four different attributes:

- Number: 1, 2, or 3 symbols
- Color: red, green, or purple symbols
- Shading: empty, striped, or solid symbols
- Shape: ovals, squiggles, or diamonds

Initially, 12 cards are placed face up on a table. Three cards form a *SET* if they are all the same or all different in *each* of the four attributes independently. Players scan the cards; the first to see a *SET* calls "*SET!*" and removes those cards, which are then replaced. The winner is the person at the end with the most *SET*s.

The authors have also developed a variation of the game making use of modular arithmetic. The End Game is played by hiding a card at the beginning of the game. At the end, when no more *SET*s can be found, the hidden card can be determined from the remaining visible cards. The goal of this game is to figure out the missing card and determine whether two cards in the layout make a *SET* with that card. This gives an idea of how mathematics can deepen one's interest in the game.

Who Did We Write the Book For?

We wrote the book for anyone with a deep curiosity about games. We hope that the exposition, the exercises, and the projects will help readers to understand the beauty of this game and the surprising connections to several broad areas of math. Interest in the game is the only prerequisite for this book,[1] but it doesn't hurt to have a healthy curiosity about mathematics as well. And the process of writing this book as a family has been meaningful to all of us: really, we wrote this book for ourselves.

The first half of this book is intended for anyone interested in mathematical aspects of the game. We do not assume any special mathematical training. As such, we avoid the definition–theorem–proof format common to math books.[2] Instead, we introduce topics with motivating questions and examples. We state general theorems on occasion, often justifying them with informal explanations or examples.

The second half of the book covers more advanced topics and will revisit some of the topics introduced earlier. We hope that these chapters will be accessible to everyone, though some additional background might enhance your understanding in certain places.

Features

Chapter 1 introduces the main ideas that the rest of the book will explore. Many questions are raised; only a few are answered in that chapter. We encourage you to read a question, then get a deck of SET cards out, play a game, and think about the question. Active reading of books that include equations is a must; like all things academic,[3] you will get out of this what you put into it. In that spirit, we hope that you

---

[1] Actually, if you don't know the game already, we'll help you become wildly interested in it.
[2] In a few instances, we just couldn't help ourselves. Proofs really are wonderful things, when sufficiently motivated.
[3] and nonacademic

have a deck and take frequent breaks for a game or two, even if it's a solitaire game. (We did the same while we wrote it, and unsurprisingly, it helped to inspire some of our lines of inquiry.)

Chapters 2–5 answer most of the questions raised in chapter 1; they should be read in order. Each introduces concepts that will be used in later chapters. For the most part, the chapters in the second half of the book can be read independently; you won't need to finish the material on combinatorics before proceeding to the material on geometry, for example.

Each chapter in this book has some exercises, and almost all have projects as well. These are intended to lead interested readers, teachers, or classes to deeper results. The authors have had substantial experience in creating and leading such projects, from the simple (worksheets in high-school classes) to the advanced (eight-week summer undergraduate research projects). For your help, in the back of the book we include very brief solutions to the exercises.

More Things to Say

We have really enjoyed writing this book. We hope it has the best of each of us in it, and we hope it will be fun for you to read, working through details as you go.

One of the major themes of the book is that asking questions and exploring ideas (which may or may not lead to answers) enhances your appreciation of the game. We've asked ourselves lots of questions, some of which have led to ideas in the book and some of which have not. This problem is not unique to this book—every author makes choices about what to include and what to omit. Paraphrasing Mark Twain, we would have written a shorter book if we'd had more time.

The topics that didn't make the cut were not losers; we just had more ideas than we could explore in this book. So, there's plenty more to do. We encourage everyone to ask more questions.[4] Come up with wacky ideas. See where they lead.

Here's how we're going to use the word "set" in this book. If we're talking about three cards, we'll write it as a "*SET*." If we're talking about

---

[4] Not a bad life strategy.

Figure P.1. Students in Madagascar during a spirited game of SET.

the manufactured game (or subgames, for example, when you take all the purple cards, so you've got only three attributes instead of four) we'll call it "SET." And, in the later chapters, when we talk about n attributes, where $n > 4$, we generally use the term "the n-attribute game." We hope it's not confusing.

Acknowledgments

We are grateful to Ethan Berkove, Jed Mihalisin, Justin Solonynka, Drew Knight Weller, Andria Gordon, Deborah Chun, and Chubbles. These people gave us tremendous feedback on all aspects of earlier drafts of the manuscript. If you see them, buy them a sandwich. We also received excellent suggestions from the anonymous reviewers, and wonderful advice from Vickie Kearn and her team at Princeton University Press.[5]

---

[5] If you see them, ask them to buy you a sandwich.

We also thank people who have contributed to our understanding of the mathematics behind the game. Anthony Forbes discovered the wonderful partition of the deck into maximal caps plus one card that we explore in chapter 9. David Eisenstat and Brian Lynch coded simulations to answer questions, some of which we didn't know we had until after they ran their simulations. Jordan Awan wrote the Cap Builder to explore maximal caps in lots of dimensions. Maureen Jackson wrote an honors thesis on SET under Liz's direction, which started this whole enterprise. Sarah Brachfeld worked on the structure of the six cards left at the end of the game. Mike Follett, Kyle Kalail, Katie Pelland, and Rob Won formed Liz's first summer research group on SET, exploring the partitions of the deck into maximal caps, and Jordan Awan, Claire Frechette, and Yumi Li continued that work in another summer of research.

Now, let's play!

SET and You

1.1 A GAME OF SET

Three students, Stefan, Emily, and Tanya, are playing SET, a game played with a special deck of cards. Each card in the game of SET has symbols characterized by four different attributes:

- Number: 1, 2, or 3 symbols
- Color: red, green, or purple symbols
- Shading: empty, striped, or solid symbols
- Shape: ovals, squiggles, or diamonds

The game is new to Tanya. Stefan is the dealer, but before he can deal the first cards, Tanya starts asking questions.

TANYA: *How many cards are in the deck?*
STEFAN: *That sounds like a math question.*
EMILY: starts counting the cards. *Give me a minute and I'll know!*
STEFAN: to Emily. *Cheater! Don't count. We can figure it out, using. . . *math!**
TANYA: *How did you do that?!*
STEFAN: *Do what?*
TANYA: *How did you speak asterisks like that?*
STEFAN: *I do not understand the question.*
EMILY: *He did it because we are in a book, obviously! We can speak all the symbols we want!* ☺

Tanya has asked the first math question that most people ask about the game, and Stefan's advice is directed to Tanya and Emily, but also

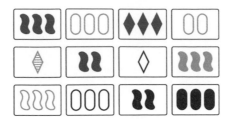

Figure 1.1. First layout of cards.

Figure 1.2. Emily's *SET*.

Figure 1.3. The *SET* not taken.

to you. Of course, if you have a deck in your hands, it's easy to answer this question the way Emily started to.

In the spirit of self-discovery, though, we will postpone answering this (and other) questions until later in the book. We encourage you to try to figure out the answers on your own. But the questions Stefan, Emily, and Tanya ask here will motivate much of what you will see in the coming chapters.

Stefan deals 12 cards—see figure 1.1.

TANYA: *How do you play the game?*
STEFAN: *You find three cards that are either all the same or all different in each of the four attributes. That's called a "SET."*
EMILY: grabbing three cards. *Like this!* (See figure 1.2.)
TANYA: *I see—all the cards have three symbols, there are three different colors, all are solid, and the three different shapes appear.*
STEFAN: *That's right! In fact, there was another SET containing one of Emily's cards, 3 Green Solid Squiggles. (See figure 1.3.)*

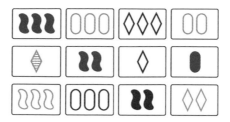

Figure 1.4. Second layout of cards.

Figure 1.5. Tanya's non-*SET*.

Figure 1.6. Tanya's actual *SET*.

STEFAN: *These are all green, but for each of the other attributes, they are all different.*

TANYA: *There were two SETs in the first layout. Is that weird?*

STEFAN: *No. There's a nice probability calculation that tells you the average number of SETs in the first layout.* (See chapter 3.) *Now, since we want to have 12 cards, I need to replace the three cards Emily took.* (Stefan deals out three more cards. See figure 1.4.)

TANYA: pointing to the three solid red cards. *Hey—is this a SET?* (See figure 1.5.)

STEFAN: *Almost! But you see there's a problem with shape: two are squiggles and one is an oval. Any time you can say "two are x and one is y," you're out of luck.*

TANYA: *Oh, now I see. How about this?* (See figure 1.6.)

EMILY: *Great! Some people find these SETs the hardest to see—all four attributes are different.*

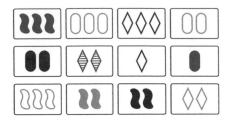

Figure 1.7. Third layout of cards.

Figure 1.8. Stefan's *SET*.

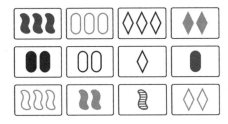

Figure 1.9. Fourth layout of cards. There are no *SET*s!

STEFAN: *Tanya's SET was actually in the original layout, but no one saw it till now.*

Tanya takes her cards and Stefan deals three more. See figure 1.7.

STEFAN: grabbing three cards. *The dealer finds a SET! That's allowed, you know.* (See figure 1.8.)

Stefan deals another three cards, and the players stare at the layout in figure 1.9 for a while.

EMILY: *I don't think there's a SET in these 12 cards. I can't find one.*
STEFAN: *Ow—my head hurts!*
TANYA: *Emily! You're hitting Stefan in the head!*
EMILY: *Sorry! Sometimes I swing my arms wildly when I'm thinking. It's pretty dangerous.*
TANYA: *Are there any SETs in here? How can we be sure?*

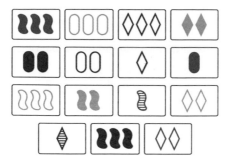

Figure 1.10. Fourth layout of cards with three cards added.

EMILY: *There aren't. There are a few ways we could check this, but for the most part, if everyone's been staring for a while and no one has found anything, we put out three more cards.*

TANYA: *How often does it happen that there are no SETs among the 12 cards?*

STEFAN: *I think that's hard to calculate. But people have estimated how often this happens by using computer simulations.*

The interlude (following chapter 5) includes methods to verify that there are no *SET*s in a given layout, and chapter 10 deals[1] with simulations. When everyone agrees there are no *SET*s, three cards are added to the layout. (See figure 1.10.)

EMILY: *Now there's a SET!*[2]

TANYA: *Is it possible for 15 cards to have no SETs?*

STEFAN: *Yes. In fact, you can have as many as 20 cards without a SET, and that's the most you can have. This turns out to be a question related to finite geometry. (See chapter 5.)*

TANYA: *Cool. So, I now know it's possible for 12 cards to have no SETs. What's the maximum number of SETs 12 cards can have?*

STEFAN: *Well, it's kind of amazing, but this is also a geometry question.* (There's a project at the end of chapter 5 devoted to constructing collections of 12 cards that have a prescribed number of *SET*s.)

---

[1] This is a pun, but it was unintentional.
[2] See if you can find one. Then, see if you can find a second one.

(a) Three attributes the same and one different.

(b) Two attributes the same and two different.

(c) One attribute the same and three different.

(d) All four attributes different.

Figure 1.11. The four different kinds of *SET*s.

TANYA: *Is it known how many SETs there are in the entire deck?*

EMILY: *Yes. That's a fun calculation.* (The answer appears in chapter 2, where lots of things get counted.)

STEFAN: *And every card is in the same number of SETs!*

TANYA: *Thanks for answering a question I didn't ask. And I can't help but notice that you aren't actually answering any of my questions. Something else I noticed: the first few SETs we found didn't always have the same number of attributes that were the same. How many different kinds of SETs are there?*

EMILY: looking through the deck. *Four. Here are examples of every possibility for how many attributes are the same.* (See figure 1.11.)

At this point, the game proceeds as before, with the players taking *SET*s and Stefan dealing more cards. After taking as many *SET*s as they can find, there are six cards left. (See figure 1.12.)

TANYA: *Are there any SETs remaining in the final layout?*

EMILY: *Nope. Game over!*

TANYA: *Is it typical for there to be six cards at the end?*

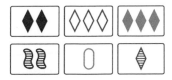

Figure 1.12. Six cards remain at the end of the game.

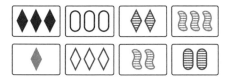

Figure 1.13. One card at the end of the game is missing—can you find it?

STEFAN: *This is a hard probability to calculate exactly, but based on computer simulations, it's usually true that either six or nine cards remain. Sometimes we can clear the deck, but that's fairly uncommon.*

EMILY: *It's also possible for 12, 15, or 18 cards to be left, but we've never actually played a game with either 15 or 18 cards left at the end.* (There are several simulations in chapter 10 that explore this topic.)

TANYA: *So always multiples of three—that makes sense. How often are there exactly three cards left?*

STEFAN *and* EMILY, together: *Never!*

At this point, the game ends, the players count their *SET*s, and Emily wins. But Tanya wonders why there can't be three cards left at the end of the game. A complete explanation uses modular arithmetic, which you will find in chapter 4.

TANYA: *That was fun! Can we play another game?*

The group plays a second game, and this time, there are eight cards left at the end of the game. (See figure 1.13.)

TANYA: *Wait! How can there be eight cards left? I thought it had to be a multiple of three! Is there a card missing or something?*

EMILY: *Yes, precisely! We hid one card at the beginning of the game, then played the usual way.*

TANYA: *Why would you do that? What is the missing card?*
STEFAN: *The amazing thing is that you can figure it out!*
TANYA: *But I didn't memorize every card that's been played!*
EMILY: *You don't need to—you can determine the missing card from the cards left on the table!*
TANYA: *How!?*

Rather than answer Tanya's impassioned plea, Emily explains what she and Stefan are doing. They (and we) call this the *End Game*.

The End Game

1. At the beginning of the game, remove one card from the deck (without looking at it!) and put it aside.
2. Now deal 12 cards face up, and play the game as usual, removing *SET*s and replacing the cards you took.
3. At the end of the game, you can determine the hidden card using just the cards left on the table.
4. Finally, now that you've determined the hidden card, you might be lucky enough to find a *SET* using the hidden card and two of the cards that are left on the table.

We'll explain how this procedure works in detail in chapter 4; we'll also discuss how often the missing card makes a *SET* in chapter 10. For now, see if you can find the missing card in the configuration in figure 1.13. [Hint: Concentrate on each attribute separately: first, determine the color of the missing card, then the number, and so on. The answer is at the end of the chapter.]

STEFAN: *OK Tanya, here's how to find the missing card.*
(∗Inaudible whispers∗)
TANYA: *Now I get it! This is so cool! The missing card is . . .* (Tanya shouts the missing card so loudly that we couldn't hear it.)
EMILY: *Perfect! It gets better. Does the missing card form a SET with two of the eight cards left on the table?*
TANYA: *Yes! In fact, it's in two different SETs.*

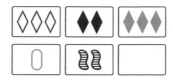

Figure 1.14. Find the missing card at the end of the game.

STEFAN: *That's right. It's really impressive when you yell "SET!," then take two of the cards on the table and finally turn over the hidden card.*

EMILY: *Yeah, it really looks like a magic trick!*

For practice (without any instruction), see if you can determine the missing card in figure 1.14. The identity of the missing card also appears at the end of this chapter.

TANYA: *Does this trick always work? Can you find the missing card if there is a different number of cards left?*

EMILY: *Yes, it always works, but you won't necessarily be able to form a SET with two of the cards on the table.*

STEFAN: *By the way, when we play the End Game, if there are five cards left, then the missing card will never form a SET with two of the cards on the table.*

TANYA: *I think I understand why. Is it related to the fact that it's impossible to have just three cards left at the end of the game?*

STEFAN: *Yes—this uses modular arithmetic.*

TANYA: *This is all so cool, and oddly foreshadowing! Let's play another game.*

Stefan, Emily, and Tanya play another game. Even though she's new to the game, Tanya does well, partly because Stefan and Emily give themselves a handicap by not immediately taking *SET*s they find. (See the interlude for some possible ways that experienced players can play with new players so that the game is fun for everyone.) When the game is over, the players have another discussion.

(a) Three cards that aren't a *SET*; call them *A*, *B*, and *C*.

(b) These three cards make *SET*s with the three pairs of cards in (a): The first card makes a *SET* with *A* and *B*, the second makes a *SET* with *A* and *C*, and the third makes a *SET* with *B* and *C*.

(c) These three cards complete *SET*s with the cards in (b).

Figure 1.15. Completing as many *SET*s as possible.

TANYA: *So, you two know so much about this game, there's gotta be another trick you can teach me.*

STEFAN: *Indeed, there is. I'm going to hand you three random cards that aren't a SET. For each pair of those cards, find the third card that makes a SET with that pair.* (Stefan puts out the three cards in figure 1.15(a).)

TANYA, hunts through the deck and finds the three cards in figure 1.15(b): *OK, these are the three cards I found.*

EMILY: *Good. Now do the same thing with the three cards you just found.*

TANYA: *Done.* (She lays down the three cards in figure 1.15(c).)

STEFAN: *Now do the same thing with the three cards you just found.*

TANYA: *You've got to be kidding. This could go on forever!*

EMILY: *It could, but it doesn't. Keep going.*

TANYA: *What is happening?!? It's the same cards we started with!*

STEFAN: *Now, look at these nine cards. Can you organize them nicely so that you can see all the SETs?*

After some reorganizing, Tanya lays out the cards in figure 1.16.

TANYA: *You're right, that was a great trick. And look at how pretty this is!*

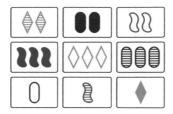

Figure 1.16. The nine cards in figure 1.15 nicely organized.

EMILY: *This is what the Set Enterprises website calls a magic square. Notice that no matter which two cards you pick, the third card that makes a SET is in there!*

We'll return to these special layouts in the next section. In the meantime, try doing this trick yourself.

1.2 MORE QUESTIONS AND A PREVIEW

In this section, we give an overview of several of the ways SET and math are related. We hope this whets your appetite for much of what follows. We will pose many questions, but just like Stefan and Emily, we will answer very few of them, at least for now. We encourage you to read actively, thinking about the questions and trying to find your own solutions. But first, a word from our sponsor.

History

SET was invented in 1974 by Marsha Falco, a population geneticist studying epilepsy in German shepherd dogs. She had a card for each dog, and she placed symbols on the card to represent that dog's expressions of various genes. As she looked at the cards, she realized she could make a game out of them. At first, she played with her family, and then in 1990, she founded Set Enterprises, Inc., to develop and market the game.

SET has repeatedly been recognized as an outstanding game, winning the TD*monthly* (ToyDirect) Top-10 Most Wanted Card Games every year from 2006 to 2015, the Mensa Select Award (1991), and the

Figure 1.17. The winner (left), Marsha Falco (right).

Parents' Choice Best 25 games of the past 25 years (2004). There has been one National SET Competition (in August 2006) advertised on the Set Enterprises' website www.setgame.com, which was won by one of the authors of this book.[3] See figure 1.17.

Counting Questions

People (including mathematicians, who are also people) started asking questions about SET as soon as it appeared in toy stores. But, as is typical in mathematics, just knowing the answer to a question is not enough. Answers often lead to more questions, and, in the case of SET, these new questions expose deeper connections between the game and math.

We begin with the first question Tanya asked.

- How many cards are needed to make the deck?

The lazy solution is to get a deck and count all the cards. You should get 81. A mathy explanation for this is the following: since there are four attributes, and each attribute has three possibilities, there are $3 \times 3 \times 3 \times 3 = 3^4 = 81$ possible cards. Why do we multiply (instead of adding, for instance)? Because we need to choose a number AND

---

[3] Just ask Hannah which one.

Figure 1.18. The fundamental theorem of SET tells you that there is a unique card that makes a *SET* with these two cards. What is it?

a color AND a shading AND a shape. Replacing AND with × is sometimes called the *multiplication principle* in textbooks on discrete math. We explain this fundamental idea more carefully in chapter 2.

- How many *SET*s are there?

A short answer: Enough to make the game interesting. We answer this question in chapter 2.

- What percentage of the *SET*s differ in all four attributes? Three attributes? Two attributes? Only one attribute?

The very clean answer to this question is explained rather carefully in chapter 2. The calculation uses some basic counting techniques. For now, you might enjoy trying to guess which kind of *SET* is most common, and which is least common.

- How many different *SET*s contain a given card?

Stefan mentioned that each card is in the same number of *SET*s. But this counting question introduces an important idea, so we'll answer it now. We'll return to it in chapter 2.

Finding the number of *SET*s that contain a given card uses a principle so important that we call it the fundamental theorem of SET.

FUNDAMENTAL THEOREM OF SET

Given any pair of cards, there is a unique card that completes a *SET* with the pair.

Two cards are shown in figure 1.18. It should be clear that there is a unique card that completes a *SET* with those two cards.

Here's how we can apply this theorem. First, choose a card C. Then the other 80 cards can be split up into 40 pairs, each of which makes a *SET* with C. This tells us that there are 40 *SET*s that contain any given card.

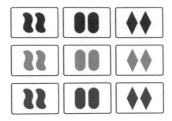

Figure 1.19. Lots of *SET*s.

By the way, we can use this fact to calculate the total number of *SET*s. Your argument might begin, "Since there are 81 cards and 40 *SET*s that contain that card, we get 81 × 40 = 3240. But this overcounts the number of *SET*s because we've counted each *SET* three times. So...."

Geometry Questions

The connection between the game and geometry is surprising. The game is played with a finite deck of cards, and standard Euclidean geometry isn't finite (there are an infinite number of points on a line, lines in a plane, and so on). But the connection to *finite* geometry is fundamental. We explore this in chapters 5 and 9. For a warm-up, try this:

- How many *SET*s can you find in the collection of nine cards in figure 1.19?

The answer is below.[4] As Emily mentioned, the Set Enterprises website calls a configuration like this a *magic square*. Unfortunately, the term magic square means something else to mathematicians.[5] It has the largest possible number of *SET*s that can be found in nine cards. We (and most mathematicians) call this configuration a *plane*, for reasons that will become clear in chapter 5.

How is this related to geometry? Think of the nine cards as "points" and the *SET*s as "lines." Then we can redraw this picture as in

---

[4] There are 12 *SET*s, although this may not be the answer to the question "How many can you find?"

[5] And to Ben Franklin, who made a study of magic squares. A *magic square* is a square array of distinct integers where each column, each row, and both main diagonals add up to the same sum. There is no relation between these magic squares and SET.

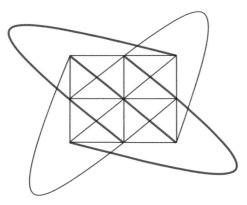

Figure 1.20. Schematic diagram of SETs in figure 1.19. This is the affine plane with
three points per line.

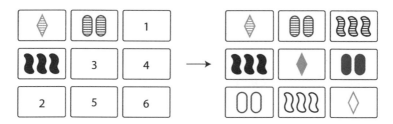

Figure 1.21. Complete the plane.

figure 1.20. (Those swoopy curves are "lines" passing through three
points, as you can see by looking at the corresponding cards in
figure 1.19. In this geometry, lines don't have to be straight!)

If you look back at the nine cards that Tanya organized in figure 1.16,
you'll see that the SETs in that figure are in exactly the same relative
positions as the SETs in figure 1.20. So her nine cards also form a plane.

In fact, you can make a plane like this from any three cards that do
not form a SET, like Tanya did (see exercise 1.3). She used a two-step
procedure, where first she found the cards and then she organized them.
You can also do this in one step, by following these simple instructions:
take three cards that aren't a SET, and put them in the corner of a
square, as in the left side of figure 1.21. The numbered spaces give you
one possible order to add cards that complete SETs. The finished result
is shown in the same figure on the right.

There are two things that seem like "magic" about planes. First, if you take any pair of cards from the plane, the third card that completes the *SET* is also in the plane. Second, if you take any three cards that aren't a *SET* from the plane, and follow the procedure in figure 1.21, you'll get the same nine cards.

We'll revisit this topic in chapter 2 when we count the number of planes in the deck, and also in chapter 5 when we investigate the game using the axioms of geometry. These planes are also good starting places for creating 12-card layouts that contain the maximum possible number of *SET*s. You're asked to do that in exercise 1.4.

There are a few observations that have a geometric flavor; these are so important, we can't wait to tell you about them:

- Given any two cards, there is a unique third card that completes a *SET* with them. (The fundamental theorem of SET.)
- Given any three cards that don't form a *SET*, there is a unique plane (up to reordering the cards) containing them.

Compare these statements with the fundamental facts you (may have) learned in high-school geometry:

- Given any two points, there is a unique line containing them.
- Given any three non-collinear points, there is a unique plane containing them.

Our geometry is different from Euclidean geometry (in particular, "lines" have only three points, and don't need to be straight). But a large and somewhat surprising amount of Euclidean geometry will apply to SET. In chapter 5, we'll learn that the SET cards form the *affine geometry* AG(4, 3), and this will have some important consequences.

There is one interesting consequence of the geometric approach:

- All *SET*s are the same.

But this is crazy. We know that there are four *different* kinds of *SET*s (see figure 1.11). However, from the geometric point of view, *SET*s are simply lines in a finite geometry, and all lines are the "same." This can be made more precise using linear algebra, which we do when we revisit this question in chapter 8.

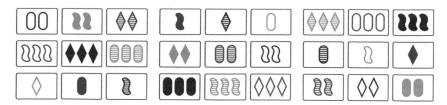

Figure 1.22. A hyperplane.

One final comment on geometry, for now. Thinking of our *SET*s as one-dimensional lines and the planes as two-dimensional objects, we are naturally led to create three-dimensional hyperplanes, as in figure 1.22. These consist of 27 cards that can be split up into three parallel two-dimensional planes. For comparison, note that the plane of figure 1.19 can be split up into three parallel lines. (Parallel *SET*s are discussed in chapters 5 and 8.)

What's special about *SET*s, planes, and hyperplanes? These collections of cards are *closed*. This means that given any pair of cards in the collection, the card that completes the *SET* with that pair is also in the collection. In chapter 8, we'll see that there are only five types of closed collections of cards: single cards, *SET*s, planes, hyperplanes, or the entire deck.

For entertainment, try to find the locations of some *SET*s in the hyperplane shown in figure 1.22. You should notice fairly quickly that if you pick any two cards, the *SET* that contains them lies within the hyperplane. As you're looking, see if you can make some sense out of the positions that the cards in a *SET* occupy in the array. If you like looking for patterns, this should offer you lots of practice.[6] We'll count the number of *SET*s in a hyperplane in chapter 6 when we consider generalizations of the game.

Finally, the entire deck of SET cards is a four-dimensional geometry, consisting of 81 points and lots of lines, planes, and hyperplanes. This will be displayed in a rather striking way in chapter 5.

---

[6] You, as a human, are wired for pattern recognition. This game, and much of mathematics, is really an elaborate pattern recognition game.

Probability and Simulations

Most games have an element of chance, because luck helps to spice things up. SET is a game of skill, but the order the cards are dealt obviously introduces uncertainty into the game. How the game unfolds also depends on which *SET*s are taken along the way, which introduces a second level of chance.

Here are a few more questions that might occur to you.

- Suppose three cards are chosen at random. What are the chances they form a *SET*?

Many probability problems are just counting problems in disguise. We'll describe two different ways to do this in chapter 3.

- Why does the game begin with 12 cards?

Well, those are the rules. But it's worth figuring out why 12 is the "right" number for playing the game. We will do so in chapter 3, when we calculate the expected number of *SET*s among 12 randomly chosen cards.

The next few probability questions are frequently asked by people who have played the game a lot. Unfortunately, they seem to be quite difficult to answer precisely. But it's possible to estimate the answers by playing the game millions of times.[7] We give the results of some simulations in chapter 10.

- What is the probability that there are no *SET*s in the initial layout of 12 cards?

Simulations indicate that this happens approximately 3.2% of the time.

- What is the probability that there are no cards left at the end?

Simulations suggest that this happens even more infrequently, approximately 1.2% of the time.

- Suppose you have a shuffled deck in your hands. Is it always possible that you could take *SET*s in such a way that there are no cards left at the end?

---

[7] Better yet, ask a computer to do this. Ask politely.

TABLE 1.1.
Assignment of coordinates to cards.

Attribute	Value		Coordinate
Number	3, 1, 2	↔	0, 1, 2
Color	green, purple, red	↔	0, 1, 2
Shading	empty, striped, solid	↔	0, 1, 2
Shape	diamonds, ovals, squiggles	↔	0, 1, 2

In playing the game many times, we can backtrack (changing the game by taking a different *SET* earlier) to get a different number of cards at the end. In fact, people who have played online games where multiple people play the same deck may have noticed that different games (with the same deck) end with different numbers of cards left on the table. Does every deck have a way to clear it? We give an answer in chapter 10.[8]

Coordinates and Modular Arithmetic

The game of SET is intimately tied to the number 3: there are 3 cards in a *SET*, $3^2 = 9$ cards in a plane, $3^3 = 27$ cards in a hyperplane, and $3^4 = 81$ cards in the deck. This connection is best understood using coordinates, where each card will have its number, color, shading, and shape specified. Converting those attributes to numbers will allow us to perform arithmetic on the deck.

We will need to encode each card as an ordered list of four numbers. We make an (arbitrary) choice in table 1.1.

Using this setup,[9] the card consisting of 3 Purple Empty Squiggles will be represented by the coordinates $(0, 1, 0, 2)$. Which card is represented by $(0, 0, 0, 0)$? It's 3 Green Empty Diamonds, which is not special in any way. This illustrates the arbitrary nature of this process, but we will stick to these assignments throughout this book.

---

[8] Skip ahead, and you'll finish the book rather quickly.

[9] If you are curious, we ordered color and shape alphabetically. That's the kind of people we are.

Figure 1.23. A *SET*.

Modular arithmetic is sometimes called *clock* arithmetic. Here's a standard problem that you may have seen before:

- It's currently 10:30 a.m. What time will it be in 100 hours?

Here's a solution. (Avert your eyes to do this yourself!) First, let's pretend that it's 10:00 a.m. (we'll add in the half hour at the end of the problem). We know 10:00 is 10 hours after midnight. In 100 hours, it will be 110 hours after (that same) midnight. Then divide by 24 and compute the remainder: since $110 = 4 \times 24 + 14$, we know it's now 14 hours after midnight (4 days later). We will write $110 = 14 \pmod{24}$.[10] So the time will be 2:30 p.m. (Alternatively, adding $100 = 4 \times 24 + 4$ hours adds 4 days and 4 hours to the current time. This also uses remainders after division by 24.)

In the clock problem, we are working mod 24 since there are 24 hours in a day. The key step to force our final answer to be a time between 0 and 23 is to first divide by 24, then find the remainder. Modular arithmetic concentrates solely on remainders.

Here's how modular arithmetic, specifically mod 3, is useful to the game of SET. Choose your favorite *SET*, which might be the one shown in figure 1.23.

What are the coordinates for the cards in this *SET*? Using our assignments from table 1.1, we get $(0, 1, 2, 1)$, $(0, 1, 2, 2)$, and $(0, 1, 2, 0)$, from left to right. What happens when we add these coordinates one at a time?

1. Adding the first coordinates (which correspond to the number of attributes on the card) gives us $0 + 0 + 0 = 0 \pmod{3}$.
2. Adding the second coordinates (which correspond to the color of the card) gives us $1 + 1 + 1 = 3 = 0 \pmod{3}$, since 3 has remainder 0 when you divide by 3.

---

[10] Math books usually use an "equals" sign with three bars here: $100 \equiv 4 \pmod{24}$. For now, we won't.

Figure 1.24. Not a *SET*.

3. Adding the third coordinates (the shading attribute) gives
$2 + 2 + 2 = 6 = 0$ (mod 3), since the remainder when you
divide 6 by 3 is also 0.
4. Adding the fourth coordinates (shape attribute) gives you
$1 + 2 + 0 = 3 = 0$ (mod 3).

So each sum is 0 (mod 3), and we get that the sum of the three cards is
just $(0, 0, 0, 0)$ (mod 3).

What happens if we do this for three cards that are not a *SET*? Try
this yourself for the three cards shown in figure 1.24.

What's the takeaway message from these two examples? It's the
following striking result:

Takeaway Message:

- Suppose A, B, and C are the vectors for three cards that form a
SET. Then $A + B + C = (0, 0, 0, 0)$ (mod 3).
- Conversely, suppose A, B, and C are the vectors for three cards
that do not form a *SET*. Then $A + B + C \neq (0, 0, 0, 0)$ (mod 3).

This is true *regardless* of how we assign our coordinates, as long as
we are consistent (and use the numbers 0, 1, and 2, and work mod 3).
Modular arithmetic will be very useful for us throughout the book.

We have one final comment about the power of modular arithmetic.
Why are the cards in figure 1.24 not a *SET*? The problem is shading:
two cards are empty, but one is solid. The sum of those coordinates
is $(0, 0, 2, 0)$ (mod 3), and the nonzero coordinate occurs in the third
spot, which corresponds to shading. This will connect SET with error-
correcting codes in chapter 8.

Advanced Topics

The second half of the book is devoted to more advanced topics. Here
is a (very) brief overview. *Affine geometry* is an important area of
mathematics, and many of its classical theorems have interpretations

through the game. Some of this will involve extending the game by adding more attributes:

- SET has four attributes: number, color, shading, and shape. What if we add more attributes to the game?

One way to play a five-attribute version of the game is to buy three decks, and mark each card in one of the decks with polka dots and each card in another deck with stripes, for example.[11] But adding attributes is an easy thing to do abstractly.

- Suppose there are $n > 4$ attributes. How many *SET*s are there? How many planes? Higher-dimensional hyperplanes?

We answer these questions in chapter 6. Considering more than four attributes will lead to general formulas, and those formulas have connections to classical counting problems.

When there are n attributes, there are n different kinds of *SET*s: all attributes different, all but one attribute different, and so on.

- How many *SET*s of each kind are there?

This is not too difficult to calculate exactly, and we can figure out which kind of *SET* is most common, and which is least common. We answer these questions in chapters 6 and 7.

Finally, there are famous unsolved problems that we can interpret in terms of the n-attribute game:

- In n-attribute SET, what is the maximum number of cards you can have with no *SET*s?

This number is known when $n \leq 6$ (at present), but not for any larger values. This question is the focus of some very high powered research, and the problem has attracted the interest of some of the top mathematicians in the world. We explore this question in chapter 9.

We love this game and its mathematics, and we hope this book motivates you to think about SET (and other games) from a mathematical perspective. Like everything in math (and the rest of life), you will understand things best by working out the details yourself.

---

[11] Warning: Actually playing this five-attribute game will give you a headache. In the head.

Figure 1.25. Exercise 1.2.

The exercises at the end of each chapter give you a chance to explore, on your own, some of the ideas we've introduced. Some are designed for practice and some lead to deeper topics that we'll return to later in the book. Enjoy!

EXERCISES

EXERCISE 1.1. The people who manufacture SET use a shorthand notation when they encode specific cards of the game as pdf files. For instance, the card with 3 Red Empty Squiggles is abbreviated 3ROS. (For some reason, they use "O" for "open" instead of "E" for "empty.") With this shorthand scheme, what card has a code that forms the basis of many Western religions?

EXERCISE 1.2. Suppose the six cards in figure 1.25 are left at the end of the game.

Here's a Stupid SET Trick:[12]

- Arbitrarily break up the six cards into three pairs; for example, you could make the pairs AB, CD, and EF.
- Figure out the three cards X, Y, and Z that complete these three pairs to make three SETs (so ABX and CDY and EFZ are all SETs).
- Then XYZ is a SET!

Try this for different ways of breaking up the six cards into three pairs (there are 15 different ways to pair them up, but you don't need to try this for all of them). [We'll see why this works in chapter 4.]

EXERCISE 1.3. There are three cards in figure 1.26. Add six cards to complete a plane.

---

[12] It's actually not stupid, it's great. Back in the day, David Letterman hosted a late night talk show that occasionally had Stupid Pet Tricks, so we borrowed the title.

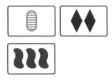

Figure 1.26. Exercise 1.3.

Figure 1.27. Build a SET ladder from Emily's *SET* to Tanya's, in order from left to right.

EXERCISE 1.4. Find a collection of 12 cards that contain 14 different *SET*s. [Hint: Start with 9 cards that form a plane, as in figure 1.19. This will reappear as part of project 5.1.]

EXERCISE 1.5. The 27 red cards in the deck form a hyperplane. How many *SET*s are there? [We'll return to this question and some generalizations in chapter 6.]

EXERCISE 1.6. *Word ladders* are games where you transform one word to another by changing one letter at a time. A standard example changes the word COLD to WARM in four steps:

$$\text{COLD} \to \text{CORD} \to \text{CARD} \to \text{WARD} \to \text{WARM}.$$

A *SET ladder* connects one *SET* to another, changing one attribute at a time. For SET, however, we'll insist that exactly one card stays the same at each step. See figure 1.11 for an example, where the *SET* at the top of the figure is changed to the *SET* at the bottom by first changing color, then shading, then shape.

a. Find a SET ladder joining the two *SET*s in figure 1.27. (The first is Emily's first *SET* from section 1.1, and the second is Tanya's.)
b. Suppose you want your favorite *SET* to be the first *SET* in a ladder. How many different *SET*s can be the next *SET* in the ladder?
c. What is the largest number of steps a SET ladder can need? Give an example of two *SET*s that achieve this maximum.
d. Change the rules! You can change the rules in any way you like, and then ask the same questions as above. For example, you could allow a color change without requiring that one card stay the same, so an all-red *SET*

Figure 1.28. Exercise 1.7.

could become an all-green *SET*, or a *SET* with all colors different could have the colors cycle through the cards. Another choice might be to allow a rearrangement of the cards. Armed with your new rules, find a *SET* ladder between the two *SET*s in figure 1.27. How many *SET*s could be the next *SET* for a particular *SET*? What's the longest distance between two *SET*s?

EXERCISE 1.7. There is a non-*SET* in figure 1.28.

a. Which attribute (or attributes) are wrong for this non-*SET*? (An attribute is "wrong" if you can say "two are one thing, while one isn't.")
b. Find the coordinates for the three cards.
c. Add the coordinates for those cards, mod 3, and call the result X. Find the coordinate positions of X that are not 0. What is the connection between those positions and the attribute (or attributes) that are wrong?
d. Do you think that it's likely that these three cards could mistakenly be taken as a *SET* during play of the game? Explain.

EXERCISE 1.8. The three cards of figure 1.28 are not a *SET*. Call these three cards A, B, and C (in left-to-right order).

a. Replace the first card A in this non-*SET* (2 Green Empty Ovals) with a card D so that BCD forms a *SET*. Find the number of attributes the cards A and D differ in.
b. Now repeat part (a) for the second card B (finding a card E so that ACE is a *SET*) and the last card C (finding a card F with ABF a *SET*). How many attributes do B and E differ in? How about C and F?
c. True/False:

   i. The 3 pairs AD, BE, and CF all differ in the same number of attributes.
   ii. The cards D, E, and F form a *SET*.

Answers to End Game questions

- Figure 1.13: The missing card is 2 Purple Striped Diamonds, and there are two *SET*s that can be formed using this card and the remaining cards on the table.
- Figure 1.14: The missing card is 1 Red Striped Diamond, and there are no *SET*s that can be formed using this card and those on the table.

Counting Fun!

2.1 INTRODUCTION

Your friends Samantha, Ethan, and Tatiana are playing SET. They are going to walk you through the delightful world of counting and the game.

ETHAN: *I am so excited to be in a book, you guys!*

TATIANA: *Me too, but what happened to Stefan, Emily, and Tanya from the first chapter?*

ETHAN: *I guess they made a SET, so they got taken away and now we're replacing them?*

SAM: *Should I worry about what's going to happen to us at the end of this chapter?*

ETHAN: *Maybe when we get replaced we'll end up in some sort of post-apocalyptic romcom buddy-cop novel!*

TATIANA: *We may just have to wait for you to write that novel. For now, we're in a book about SET.*

SAM: *And it's a good thing, because I was intrigued by some of the counting questions that were posed in the first chapter.*

TATIANA: *Like what?*

SAM: *Well, how do we count the number of SETs, and the number of SETs of each kind? How do we know anything? I have so many questions!*

ETHAN: *Well, once we know how to count the number of SETs in the deck, we can use that technique to figure out loads of things, like how many SETs there are of each kind, probabilities, and a whole bunch of other things related to advanced math topics. So let's get going!*

2.2 BASIC COUNTING QUESTIONS

Combinatorics is the branch of mathematics that deals with counting. The name "combinatorics" comes from the word "combination," which we'll see soon. Counting questions are found throughout mathematics, and the techniques we develop here can be used in many other contexts. To get started, let's look at the kind of question you might see in a book.[1] Sam asked several questions, so here's the first.

Question 1: What is the total number of cards in the deck?

As mentioned in chapter 1, there are four attributes, and each attribute has three expressions. But why does this mean that there are $3 \times 3 \times 3 \times 3 = 3^4$ cards? This is a very nice application of one of the most fundamental counting principles in combinatorics, the multiplication principle (sometimes called the fundamental counting principle).

The key to the solution is to realize that there is the same number of cards for each attribute expression. Looking at color, for example, we see the number of red cards has to be the same as the number of green cards and the number of purple cards. Similarly, for shapes, there are as many cards with diamonds as there are with ovals or with squiggles, and the same holds for the number and shading attributes.

So, begin with one attribute, color, and just focus on the red cards. Now the red cards can be broken up by shape, so pick a shape, say diamonds. How many cards in the deck have red diamonds? You can see them all in figure 2.1. There are three red diamond cards with one symbol: an empty, a striped, and a solid. How about red diamond cards with two symbols? Red diamond cards with three symbols? Of course, there are three of each as well, so there are $3 \times 3 = 9$ cards with red diamonds.

---

[1] Like this one?

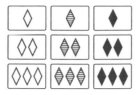

Figure 2.1. All the cards with red diamonds.

Now, there should be nine cards with red squiggles and nine with red ovals, so there are $9 \times 3 = 27$ red cards. Finally, to include the green and purple cards, we get $27 \times 3 = 81$ cards in the deck.

As an example of the multiplication principle, consider the number of different ways we could create a sentence using the following words:

My _____ _____ is _____ _____ .
        1   2   3   4

You fill in the four blanks with words chosen from the following columns:

1	2	3	4
tall	dentist	usually	confused
friendly	neighbor	not	obnoxious
disgruntled	aunt	often	screaming

How many different sentences can you create this way? Well, if you can use any of the choices from the columns, there are $3 \times 3 \times 3 \times 3 = 81$ possible sentences. For instance, "My tall neighbor is not screaming" is one of these 81 possibilities.

Replace the table we used for our sentence creator with another table:

	1	2	3	4
1		green	empty	diamonds
2		purple	striped	ovals
3		red	solid	squiggles

It should now be clear why there are 81 cards in the deck. We turn to Sam's first question.

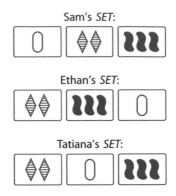

Figure 2.2. These three *SET*s are the same *SET*.

Question 2: How many *SET*s are there in the 81 cards?

This uses another counting technique from combinatorics. To compute the total number of *SET*s, we need to understand *permutations* and *combinations*. For example, let's say Sam, Ethan, and Tatiana had 12 friends over, and they wanted to pick two friends to bring into a secret room for cake. First picking Tiana, and then picking Silvia, isn't any different from picking Silvia first and Tiana second, since the result would be that both Silvia and Tiana get to eat cake. This is a *combination*: order doesn't matter. However, if the first person they choose gets the whole cake and the second person gets just a small cupcake, then it *would* matter whether Silvia or Tiana were chosen first. That's a *permutation*.[2]

A. When order doesn't matter, and we are just choosing groups (or subsets) of people or things, we're looking at *combinations*.

B. When order does matter, and we are ranking people or things, we're looking at *permutations*.

To figure out the total number of *SET*s, order doesn't matter. Here's why: Suppose Sam, Ethan, and Tatiana each create a *SET* in their mind,[3] as shown in figure 2.2.

Clearly, it's the same *SET*, listed three times, so order doesn't matter in creating *SET*s. How many different Sams, Ethans, and Tatianas could

---

[2] Whoever is chosen second will think this permutation is unfair.

[3] You might call this a mindset.

have picked the same *SET* in different orders? There are three choices for the first card, two for the second, and the last is determined. This is a permutation of the three cards, giving us $3 \times 2 \times 1 = 6$ ways to order the cards.

Now we're ready to count the number of *SET*s in the whole deck. This will be a combination—order doesn't matter. We first count the number of ways to choose the three cards in order, then divide by the number of ways you could have chosen those three cards, just like Sam, Ethan, and Tatiana did above.

1. How many choices do we have for the first card in the *SET*? We have 81 choices since there are 81 cards in the deck.

2. Once we've chosen that card, how many choices do we have for the second card? Since we've removed one card, there are 80 choices.

3. How about the third card? The fundamental theorem of SET (FTS) tells us that, given any two cards, there is a unique card that completes the *SET*. That means that there's only one card left to complete our *SET*.

So the total number of ways of choosing three cards in order that make a *SET* is $81 \times 80 \times 1 = 6480$. But this isn't the number of *SET*s because, as Sam, Ethan, and Tatiana have shown us, this procedure counts each *SET* 6 times. Then the total number of *SET*s is

$$\frac{81 \times 80 \times 1}{3 \times 2 \times 1} = 1080.$$

Now that we have the total number of *SET*s, we'll conclude this section with one final basic counting question, which we'll actually answer twice.

Question 3: How many *SET*s contain a given card?

Pick a card, any card. Once you've selected a card, there are 80 cards left, and if you pick one of them, then by the FTS, there is a unique third card that completes a *SET* with those two. So that means the 80 cards come in $\frac{80}{2} = 40$ pairs, where each such pair forms a *SET* with our original card.

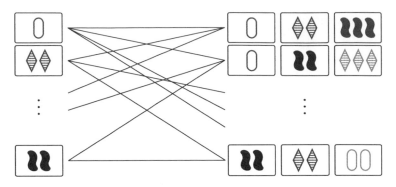

Figure 2.3. A relationship between cards and *SET*s.

ANOTHER WAY TO COMPUTE HOW MANY *SETS* CONTAIN A GIVEN CARD

Mathematicians often like to count the same thing twice, because counting something in two different ways can give you new formulas[4] and answers to new questions. We will count the number of *SET*s containing a given card using *incidence counting,* a nice way to illustrate combinatorial reasoning. We'll use this technique in other places in this book, especially for more complicated counts.

To begin, we're going to set up an incidence graph, a type of picture with objects on the left side representing one group of items, and objects on the right side representing another group of items. We'll then draw lines joining the objects on one side to the objects on the other to represent a relationship between the two kinds of objects.

Our incidence graph is pictured in figure 2.3. On the left, we have listed all 81 cards.[5] On the right, we have all 1080 *SET*s. We draw a line from a card on the left to a *SET* on the right if the card is in the *SET*. Then each *SET* on the right is paired with three cards on the left. Each card, however, is paired with a temporarily unknown number of *SET*s (because we're pretending we don't know how many *SET*s a given card is in). Call this unknown x.

Here's what makes this beautiful: the number of lines connecting the two sides is clearly the same, regardless of how you count them. Looking

---

[4] Or "formulae," if you're into Latin.

[5] No, we haven't. But you can imagine we did. Mathematicians like to use ellipses (...) in order to avoid spending a lot of time writing very long lists.

at the right, we have 1080 *SET*s with 3 lines arriving at each one, so there are 1080 × 3 lines. On the other hand, on the left, there are 81 cards, and each card has x lines leaving it. So, the number of lines is 81 × x. But the number of lines is the same:

$$81 \times x = 1080 \times 3,$$

$$x = 40.$$

Again, we have shown that each card is in 40 *SET*s. More importantly, we have another useful counting tool.

2.3 SLIGHTLY MORE ADVANCED COUNTING QUESTIONS

SAM: *We're so clever, we've figured out so many things!*
ETHAN: *I think the authors helped. A lot.*
TATIANA: *Speak for yourself. Anyway, I'm still wondering about some of the other counting questions. We have 1080 SETs, right? How many of those have exactly three attributes the same and one attribute different? What about the other possibilities?*

Tatiana's question is going to be very important in chapter 3, when we look at probability, so let's tackle it now.

First of all, some calculations are going to be so useful that they get their own special names and notation. The number of ways to order three things is "3 *factorial*," which is written with an exclamation mark:[6]

$$3! = 3 \times 2 \times 1.$$

We used this in the previous section when we counted the number of *SET*s in the deck: we divided by 3! = 6 because there were 3! ways to order three cards.

Factorials are used in other important formulas in combinatorics. For instance, the total number of ways to choose three cards (maybe a

---

[6] See, math is exciting!

SET, maybe not) from the 81 cards in the deck is a combination, because order doesn't matter. We say "81 *choose* 3" and write $\binom{81}{3}$. The idea is the same as before: There are 81 choices for the first card, 80 choices for the second, and 79 for the third. But we've counted each choice of three cards 3! times, just as before. So,

$$\binom{81}{3} = \frac{81 \times 80 \times 79}{3 \times 2 \times 1} = 85,320.$$

There's a general formula for *n* choose *k*:

$$\binom{n}{k} = \frac{n!}{k! \times (n-k)!}.$$

Let's use the formula to calculate $\binom{81}{3}$ to show that this formula is the same as what we did before:

$$\binom{81}{3} = \frac{81!}{3! \times (81-3)!} = \frac{81!}{3! \times 78!}$$

$$= \frac{81 \times 80 \times 79 \times \cancel{78} \times \cancel{77} \times \cdots \times \cancel{2} \times \cancel{1}}{3 \times 2 \times 1 \times \cancel{78} \times \cancel{77} \times \cdots \times \cancel{2} \times \cancel{1}}$$

$$= \frac{81 \times 80 \times 79}{3 \times 2 \times 1}$$

$$= 85,320.$$

Notice how the cancelation did the trick. This formula is going to be very useful in answering Tatiana's question.

Question 4: How many *SET*s have 1, 2, 3, or all 4 attributes different?

Tatiana's question is really four questions: we want four numbers, one for each of the four possibilities, and those four numbers should add up to 1080, the total number of *SET*s in the deck.

Figure 2.4. Old standby: 1 Red Empty Oval.

We will need a more convenient way to describe the cards in the deck. Here's one: Give each of the 81 SET cards four coordinates, one for each attribute. (You might be famiilar with coordinates from the two-dimensional Cartesian plane, as in (x, y). This is the same idea, bumped up to four dimensions, as in (x, y, z, w).) Each variable will represent one of the four attributes:

- number (N = 1, 2, or 3);
- color (C = G, P, R);
- shading (Shd = E, St, So); and
- shape (Shp = D, O, Sq).[7]

Every card in the deck can now be written in its own unique (N, C, Shd, Shp) coordinates. The coordinates for the card pictured in figure 2.4 are (1, R, E, O).

In order to find the number of *SET*s that differ in a given number of attributes, we will start by choosing one card; we will then choose a second card to determine the *SET*, making sure these two cards differ in the correct number of attributes. As before, we have 81 cards to choose from at first. This means that all our calculations will begin with $81 \times x$, where x will count whatever additional options we need. Now let's narrow down those options by actually picking a card. Since Sam, Ethan, and Tatiana have been very attached to the red cards, let's choose 1 Red Empty Oval as our initial card.

Now let's count!

ALL FOUR ATTRIBUTES ARE DIFFERENT

We have picked (1, R, E, O) as our first card. In choosing a second card, we will need to eliminate some possibilities, namely every card with one symbol, every red card, every empty card, and every card containing

---

[7] Perhaps it's awkward that five different things begin with the letter S. Thanks, English!

ovals. That leaves us two choices for number *and* two choices for color *and* two for shading *and* two for shape:

Attribute:	(N,	C,	Shd,	Shp)
	↓	↓	↓	↓
Number of choices:	2 ×	2 ×	2 ×	2.

So, the number of ways of choosing two cards *in order* that share no attributes is $81 \times 2 \times 2 \times 2 \times 2 = 1296$. (Note that we've used our multiplication principle here!) Now, to finish, we need to invoke the fundamental theorem of SET: choosing those two cards has automatically chosen a *SET*. Even more, the third card must be different from each of the other two in every attribute, because that's the rule for *SET*s.

TATIANA: *That number is too big. It's bigger than the number of SETs.*

SAM: *It's OK. We need to remember that every SET got counted more than once.*

TATIANA: *Like when we each picked the same SET but in different orders!*

ETHAN: *Exactly! So we are going to need to divide that number by 3! = 6 to take care of the overcounting.*

Ethan is right: we need to remember that any time we count the number of ways to choose *SET*s in any form, we *must* divide by $3! = 6$ to get rid of repeated *SET*s. So the total number of *SET*s with all attributes different is

$$\frac{81 \times (2 \times 2 \times 2 \times 2)}{6} = 216.$$

ONE ATTRIBUTE IS THE SAME AND THREE ARE DIFFERENT

We can use the same idea as above, but this time we'll keep one attribute the same. We will pick an attribute to stay the same, then see what would have happened if we had picked another. Thinking about our Old Standby, (1, R, E, O), we'll keep the number on the card the same but change all of the other attributes. Since Old Standby has 1 symbol, the card we choose *must* also have 1 symbol. This means that there's only one choice for number; next, in order for the others to be different, there are two options for each of color, shading, and shape. We also need to remember that once we've chosen those two cards, the third card that makes a *SET* with them is determined (FTS!), and it must also have 1 symbol, and everything else will be different:

$$
\begin{array}{cccc}
(\text{N}, & \text{C}, & \text{Shd}, & \text{Shp}) \\
\downarrow & \downarrow & \downarrow & \downarrow \\
1 \ \times & 2 \ \times & 2 \ \times & 2.
\end{array}
$$

This gives $81 \times 1 \times 2 \times 2 \times 2$ ways to choose three cards that all share the same first attribute. But what if we choose to keep color the same instead of number? Then we'd have the diagram below:

$$
\begin{array}{cccc}
(\text{N}, & \text{C}, & \text{Shd}, & \text{Shp}) \\
\downarrow & \downarrow & \downarrow & \downarrow \\
2 \ \times & 1 \ \times & 2 \ \times & 2.
\end{array}
$$

There are four attributes to choose from, and we're choosing only one to remain the same. That sounds familiar...:

$$
\binom{4}{1} = \frac{4!}{1! \times (4-1)!} = 4.
$$

We'll then need to multiply $81 \times (1 \times 2 \times 2 \times 2)$ by $\binom{4}{1} = 4$, since the numbers are the same (but moved around) for each choice of the attribute that's the same. And, as above, we need to divide by 3! to account for the different orders. So the number of *SET*s with one attribute the same is

$$
\frac{81 \times (1 \times 2 \times 2 \times 2) \times \binom{4}{1}}{6} = 432.
$$

TWO ATTRIBUTES ARE THE SAME AND TWO ARE DIFFERENT

The main ideas for this case are the same as the case we just finished. We have 81 choices for the first card, then we want to choose a second card with two attributes that are the same and two that are different. We have $\binom{4}{2}$ ways of choosing those two attributes, and each choice

TABLE 2.1.
Final counts (including percentage of total deck).

Kind of SET	Number	Fraction	Percentage
All different	216	$\frac{216}{1080}$	20%
Three different	432	$\frac{432}{1080}$	40%
Two different	324	$\frac{324}{1080}$	30%
One different	108	$\frac{108}{1080}$	10%
Total	1080	$\frac{1080}{1080}$	100%

results in $81 \times 1 \times 1 \times 2 \times 2$ ordered *SET*s. Then, FTS tells us that those two cards have determined a *SET* with exactly two attributes the same. Finally, dividing by 3! gives the total number of *SET*s with exactly two attributes the same:

$$\frac{81 \times (1 \times 1 \times 2 \times 2) \times \binom{4}{2}}{6} = 324.$$

THREE ATTRIBUTES ARE THE SAME AND ONE IS DIFFERENT

Once again, the reasoning is very similar to the previous two cases. The total number of *SET*s with three attributes the same is

$$\frac{81 \times (1 \times 1 \times 1 \times 2) \times \binom{4}{3}}{6} = 108.$$

By the way, we could have cheated for this final case, but instead of thinking of it as cheating, we'll think of it as checking our answers. We know that there are 1080 *SET*s, and we know the answers for the first three parts. That means that the number of *SET*s with three attributes the same should be

$$1080 - (216 + 432 + 324) = 108.$$

We get 108, which agrees with the calculation above. (That's good.) Our results are summarized in table 2.1. We include the percentage of the total deck each kind of *SET* represents, as a preview of chapter 3.

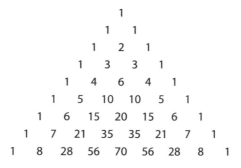

Figure 2.5. Pascal's triangle.

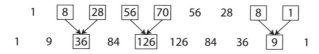

Figure 2.6. How to generate the ninth row of Pascal's triangle from the eighth.

2.4 PASCAL'S TRIANGLE: A DETOUR

TATIANA: *I'm super into this n choose k stuff.*

ETHAN: *That's just because you asked the questions.*

SAM: *But I can see why Tatiana likes it—it's really neat. I wonder where else we can use these numbers.*

We agree with Sam, and it's high time we looked at Pascal's triangle. As shown in figure 2.5, each row begins and ends with a 1, and the middle numbers in a row are created from the previous row by adding the two numbers diagonally above. For example, figure 2.6 shows how to get several entries in the ninth row.

TATIANA: *I see how to make the triangle but I don't see how it has anything to do with my awesome $\binom{n}{k}$.*

SAM: *Well, let's look at the numbers in the triangle. If the triangle got introduced in response to your questions, I'm sure there's a connection.*

ETHAN: *OK, we know how to compute combinations, so let's do some calculations and see what happens.*

Ethan has a good idea, so let's calculate a few numbers $\binom{9}{k}$:

$$\binom{9}{1} = \frac{9!}{1! \times (9-1)!} = 9,$$

$$\binom{9}{2} = \frac{9!}{2! \times (9-2)!} = 36,$$

$$\binom{9}{3} = \frac{9!}{3! \times (9-3)!} = 84.$$

TATIANA: *I see it—I see the pattern! The number in Pascal's triangle in row n and position k is $\binom{n}{k}$, if you ignore the 1s.*

SAM: *Yeah, what about the 1s at the beginning and end of each row?*

ETHAN: *And why does the row you're calling row 9 have 10 numbers in it?*

The procedure in figure 2.6 is called *Pascal's recursion*:

$$\binom{n}{k} = \binom{n-1}{k-1} + \binom{n-1}{k}.$$

This works in the triangle (as we've drawn it) only if $1 < k < n$, because the two 1s at the ends of each row don't have two numbers above them. Those two numbers represent $\binom{n}{0}$ and $\binom{n}{n}$. So apparently $\binom{n}{0}$ and $\binom{n}{n}$ equal 1, but why? If you think about it, there's only one way to choose *nothing* from 9 things, and only one way to choose all 9 things from 9 things, so we expect $\binom{9}{0} = \binom{9}{9} = 1$. Now, look at our formula for $\binom{n}{k}$ where $k = 0$ and where $k = n$:

$$\binom{n}{0} = \frac{n!}{0! \times (n-0)!} = \frac{n!}{0! \times n!} \quad \text{and}$$

$$\binom{n}{n} = \frac{n!}{n! \times (n-n)!} = \frac{n!}{n! \times 0!}.$$

To ensure both formulas give an answer of 1 (which is the correct answer), we need $0! = 1$. Fortunately, the rest of the mathematical world agrees with us:

$$0! = 1.$$

Again, you can justify this by arguing that there's only one way to order nothing.[8]

ETHAN: *So the first term of every row corresponds to* $k = 0$.
SAM: *Yeah, and we can rewrite every row using combinations, like this:*

$$\binom{n}{0} \quad \binom{n}{1} \quad \binom{n}{2} \quad \cdots \quad \binom{n}{n-2} \quad \binom{n}{n-1} \quad \binom{n}{n}.$$

TATIANA: *I love it! This means that instead of writing out every row of Pascal's triangle to get to the eighth term of the fifteenth row, all I have to do is figure out* $\binom{15}{8}$.
SAM: *As long as you remember that the eighth term is in the ninth spot.*
TATIANA: *Are you trying to confuse me?*
SAM: *No, 0 is trying to confuse you!*

Pascal's triangle is amazing, and it's useful for lots of counting and probability problems. We'll return to these numbers in chapter 6.

2.5 ADVANCED COUNTING QUESTIONS

SAM: *I really like counting. Can we try some more advanced questions?*
ETHAN: *Yeah, like how many intersets are there? And how many intersets contain a given card?*
TATIANA: *And the number of planes!*

---

[8] Another explanation is that a lot of math would break if $0! \neq 1$.

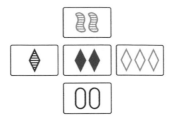

Figure 2.7. Two *SET*s that include the 2 Red Solid Diamonds card.

SAM: *Those might be good questions, but I don't know what "intersets" are, and how do we define a "plane" in reference to* SET?

There are interesting counting questions here. We begin by addressing Ethan's question about *intersets*.

Intersets

To define an interset, we first build two *SET*s that have a card in common. See figure 2.7.

If we think of the cards as *points* and the *SET*s as *lines*, then these two *SET*s pictured above are a pair of intersecting lines with the central card in common. (This geometric approach will be studied in chapter 5.)

Now remove the center card—this is how we define an *interset*: an interset is a group of four cards that can be paired so that the missing card is the same, i.e., two *SET*s that intersect in a common card, with that common card removed. We will call the missing card the "center" of the interset.

SAM: *Oh, cute—I get it. It's like "intersect." So now I know what an interset is . . . maybe. How do you know that an interset we construct with one card as the center isn't the same as an interset with a different center card?*

TATIANA: *Are you asking whether we can take one of our intersets that we constructed around the 2 Red Solid Diamonds and rearrange the 4 cards so that the center card is not the 2 Red Solid Diamonds? That's a good question.*

ETHAN, scrunches his face up as he thinks: *I figured it out! You can do it with coordinates and modular arithmetic, like*

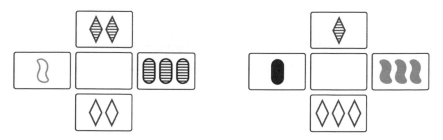

Figure 2.8. Two different intersets, each intersecting at 2 Red Solid Diamonds.

> *chapter 1 talked about.* (You can try it yourself in
> exercise 2.4.)

SAM: *Why are intersets important?*

ETHAN: *When we play the game, we often find ourselves noticing
intersets, because they aren't SETs but they are real
patterny. When we see too many of them, we guess that
there are likely to be fewer SETs.* (In chapter 3, we will find
the expected number of intersets in a 12-card layout.)

In figure 2.8, we give two more intersets with the same center as the
interset in figure 2.7, 2 Red Solid Diamonds. These figures suggest an
approach we can take to answer Ethan's questions. However, we'll start
with a question he didn't ask, because we'll use it to answer the other
two questions.

Question 5: How many intersets can we construct using a given
card as the center?

To answer this question, we'll use the answer to question 3: every
card is in 40 *SET*s. For example, take 2 Red Solid Diamonds. Because an
interset is *two SETs* built around the center card, we choose two *SET*s
that contain this card. So, the number of intersets with a given card as
the center is

$$\binom{40}{2} = 780.$$

We can use the answer to this question to address the next one.

Question 6: How many intersets are there?

Since there are 81 cards in the deck, and we just determined that each card is the center of 780 intersets, we get the total number of intersets in the deck is

$$780 \times 81 = 63,180.$$

SAM: *Wait, that's huge—much larger than the 1080 SETs we found earlier. Did we forget to divide by something?*

TATIANA: *No. One thing I've learned in my life is that math doesn't lie.*

ETHAN: *But this kind of makes sense. Asking how many intersets there are is the same as asking how many pairs of SETs there are containing a given card. And there are tons of those.*

We're now ready for Ethan's final question.

Question 7: How many intersets contain a given card?

This is different from the first count, because the central card is *not* in the interset. To do this, we set up an incidence count, as we did before.

First, we draw a graph[9] with the 81 cards on the left side, and all of the 63,180 intersets on the right. A line joins a card to an interset if the card is in the interset.

How many lines are there? On the left, the number of lines is 81 (cards) times the number of intersets each card is in, which we'll call x. On the right, the number of lines is equal to 63,180 (the number of intersets) times 4 (the number of cards in any interset). As before, the number of lines is the same, so

$$81 \times x = 63,180 \times 4.$$

Solving for x gives us the number of intersets containing a given card:

$$x = 3120.$$

---

[9] We will draw the graph in our heads, because we can. And so can you!

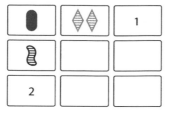

Figure 2.9. Ready to start making a plane, adding card 1 to complete a horizontal *SET* and card 2 to complete a vertical *SET*.

ETHAN: *Wow. That's a lot of intersets that contain my favorite card.*

SAM: *It makes sense, though, since there are so many intersets to begin with.*

ETHAN: *I'm curious to see what else we can do with intersets later on!*

TATIANA: *Me too. But I also asked about counting planes! Can we talk about planes now?*

Planes

In section 1.2, we defined planes as special groups of nine cards with the following property: given any pair of cards, the plane also contains the third card that the fundamental theorem of SET guarantees will make a *SET* with the pair. We also constructed a plane from three cards that were not a *SET*. We'll do so again, more carefully, so you can see that the three cards completely determine the plane.

Take three cards that are *not* a *SET* and arrange them in the upper left corner of a rectangle, as in figure 2.9. The cards labeled 1 and 2 indicate where we start completing *SET*s.

What card should we put in the box labeled 1 to complete the horizontal *SET* with the other cards in the first row? 3 Purple Empty Squiggles. What card belongs in box 2 to complete the vertical *SET* that occupies the first column? 1 Green Empty Diamond. See figure 2.10.

Now the card in box 3 completes the diagonal *SET* with the two cards 1 Green Empty Diamond and 3 Purple Empty Squiggles—2 Red Empty Ovals, as in figure 2.11.

Finally, the remaining three positions are filled by cards that complete additional horizontal, vertical, and diagonal *SET*s. What order should they be added in? It doesn't matter! Although the numbers 4, 5,

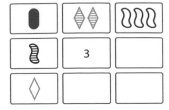

Figure 2.10. Progress on the plane; next, we'll add the card labeled 3.

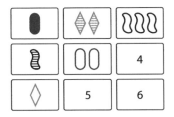

Figure 2.11. More progress on the plane; next, we'll add the last three cards.

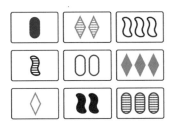

Figure 2.12. The completed plane.

and 6 seem to imply that's the order you should use, you don't have to. That's what's so wonderful about planes. See figure 2.12 for the finished product.

How many *SET*s does the plane contain? Figure 2.13 gives a hint, if you've forgotten the answer that was given in chapter 1.

Each plane contains 12 *SET*s. Now that we know what a plane is, we can count them (and finally answer Tatiana's question).

Question 8: How many planes are there in the deck?

Recall that we started our plane with three cards that weren't a *SET*. Once those three cards are chosen, the plane is completely determined. We have 81 choices for the first card, and 80 for the second card. For the third card, we need a card that does not complete the *SET* determined

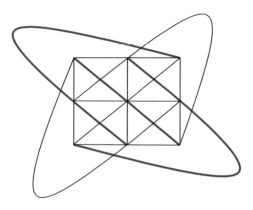

Figure 2.13. The affine plane. We call this AG(2, 3) in chapter 5.

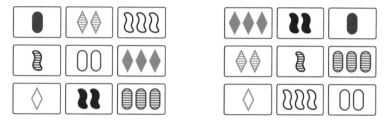

Figure 2.14. Are these two planes the same?

by the first two cards, leaving 78 valid choices. Multiplying, we get $81 \times 80 \times 78 = 505{,}440$ ways to lay out the nine cards.

Have we overcounted? It depends on whether order matters: Is one plane "the same" as another if it consists of the same cards, rearranged? For example, compare the cards in the two planes in figure 2.14.

We consider these planes to be the same.[10] When we think about *SET*s, we don't care about the order of the three cards. The special property that planes have doesn't depend on the order of the cards in the plane. While we like the pretty picture of the plane, it's the actual cards that are important.

To get an answer that doesn't depend on where the cards were placed, we need to divide by the number of ways to rearrange the nine cards. We have 9 choices for the upper left position, then 8 choices for the upper middle position. For the left middle position, we have

[10] So does the rest of the mathematical world.

6 choices (since this card cannot be the card that completes a *SET* with the first two cards). Once we have chosen those three cards, the rest of the plane is determined, as before, so we get $9 \times 8 \times 6 = 432$ ways to rearrange the cards in the plane.

This means that the total number of planes you can make from SET cards is

$$\frac{505,440}{432} = 1170.$$

TATIANA: *I love planes!*

ETHAN: *We know.*

SAM: *That was so much fun! I can't wait to see how all of our counts get used in the rest of the book!*

ETHAN: *Yeah, we did a really good job.*

TATIANA: *I think we should go celebrate with a rousing game of SET. Everyone else can play with some exercises.*

EXERCISES

EXERCISE 2.1. (Restricted counts) We can use the ideas in this chapter to count various quantities for a special subset of the cards.

a. How many red cards are there?

b. How many red *SET*s are there?

c. Given a (red) card, how many red *SET*s contain your card?

d. Among the red cards, *SET*s can have one, two, or three attributes that differ. How many red *SET*s are there of each kind?

e. How many red intersets are there?

f. How many red intersets contain a given (red) card?

g. How many red planes are there?

h. How many red planes are there that contain a given (red) card?

EXERCISE 2.2. Your friend Stefano has made a game of five-attribute SET where each card has an extra attribute: number, color, shading, shape, and feel, where each card feels pointy, slimy, or moving.[11]

---

[11] Simpsons fans might recognize this from Episode 1F06: "Boy-Scoutz 'N the Hood."

a. Use the multiplication principle to figure out the number of cards in Stefano's deck (ignore the fact that some cards may stick together).
b. How many *SET*s are there in Stefano's game?
c. How many *SET*s of each possible kind are there?
d. How many *SET*s contain a given card?
e. How many intersets are there?
f. How many planes are there?
g. Comment on how much fun playing this game might be.

EXERCISE 2.3. Edna thinks Stefano's version in exercise 2.2 is lame. She prefers her new version of SET with four expressions for each of the attributes:

- Number: 1, 2, 3, or 4
- Color: red, green, purple, or brown
- Shading: empty, striped, checkered, or solid
- Shape: ovals, squiggles, diamonds, or rectangles

a. How many cards would be in Edna's deck?
b. Assume a "*SET*" is now defined as a collection of four cards where, for each attribute, everything is the same or everything is different. Show that two "*SET*s" can now intersect in more than one card.

EXERCISE 2.4. We claimed in section 2.5 that the center of an interset is unique. Justify that this is true by first creating an interset with four cards *A*, *B*, *C*, and *D*. Then find the cards that complete *SET*s with each of the six pairs of cards *AB*, *AC*, ... Then show that this produces five different cards, so only one grouping of the four cards into two pairs will produce an interset. (A proof that the center is unique, using modular arithmetic, appears in exercise 4.6.)

EXERCISE 2.5. We counted the number of planes in the deck—there are 1170. How many planes contain a given card? [Hint: Use an incidence count!]

EXERCISE 2.6. Every plane contains 12 *SET*s (as we saw in figure 2.12). How many planes contain a given *SET*? For this problem, pick a specific *SET* and count the number of different planes that include that *SET*. [Hint: Set up an incidence count with *SET*s on one side and planes on the other.]

PROJECTS

PROJECT 2.1. (Color-blind SET!) Several of us have had the experience of playing SET with people who can't fully participate because they are color-blind and have trouble distinguishing some of the cards. First, our advice is to get a deck and put large dots in the upper left and lower right for the red cards, and large dots in the upper right and lower left for the green cards, which will allow anyone to distinguish the three colors of cards. (If you like, you can make the dots red and green. Some people say that the distraction makes the game harder for non-color-blind people, but our response is to remind those people that the game will be harder for a color-blind person anyway, so the penalty is a small one to pay.) For the sake of this project, however, suppose two people with the same kind of color-blindness want to play and have access to an undecorated deck only. How many SETs are there? How many SETs of each kind are there? In the problems below, you'll get to do these counts for various kinds of color-blindness.

a. Total color-blindness. In this condition, all colors are the same. This is as if you had a mono-colored deck, so there are only three attributes, and every card appears three times. That means you could have a SET with the same card three times.

   i. How many SETs are there? Show that it's possible for two SETs to intersect in more than one card.
   ii. How many SETs are there with all three of the distinguishable attributes different?
   iii. How many SETs are there with two of the three distinguishable attributes different and one the same?
   iv. How many SETs are there with one of the three distinguishable attributes different and two the same?
   v. How many SETs are there with all the cards the same?

b. Red–green color-blindness. Some people can't distinguish the red and green cards, but find the purple ones distinguishable from the others.[12]

   i. How many SETs are there?
   ii. Describe the different kinds of SETs that can occur, and count as many of these as you can.

---

[12] There's an app for tablets called "SET Pro HD" that has a "yellow, red, and black" color option, which may be useful for people with this type of color-blindness.

c. We've played with people who find the solid cards distinguishable, but have trouble determining the color of the striped or empty cards. For example, they are able to distinguish 1 Red Solid Squiggle from 1 Green Solid Squiggle, but they cannot distinguish 2 Purple Empty Squiggles from 2 Green Empty Squiggles.

   i. What are the color/shading possibilities for *SET*s in this case?

  ii. How many *SET*s are there?

 iii. Describe the different kinds of *SET*s that can occur, and count as many of these as you can.

Probability!

3.1 INTRODUCTION

Sam, Ethan, and Tatiana have run off to play SET, using their new counting skills. They've been replaced by a new set[1] of friends: Sophie, Eduardo, and Teddy. These friends have some new questions that are related to the counting questions we answered in chapter 2.

SOPHIE: *We've learned a lot about counting things in* SET. *But I'm wondering, now that we know the number of SETs with no attributes in common, for example, what's the probability that a randomly chosen SET has no attributes in common? What about all the other kinds of SETs? What about intersets? Planes? Can we figure out probabilities for these, too?*

EDUARDO: *Wow, you're not wasting any time. We didn't even get a chance to make inane chitchat!*

TEDDY: *Before we jump into this, we need to figure out how probability is calculated.*

EDUARDO: *I think I remember from math class that we can measure probability in percentages. If I roll a fair die, there is a 50% chance of rolling an odd number.*

SOPHIE: *No, a probability is a decimal between 0 and 1, so you say the probability of rolling an odd number is 0.5.*

TEDDY: *Actually, you're both right!*

---

[1] Isn't that clever? (This is a rhetorical question.)

3.2 WHAT IS PROBABILITY?

Let's start with an intuitive description. The probability of an event happening is the number of different ways the event can happen divided by the total number of possibilities:

$$P(\text{event}) = \frac{\text{number of ways event occurs}}{\text{total number of possibilities}}.$$

This means that a probability is a fraction between 0 and 1. Then, if you like, those fractions can be converted to decimals or percentages.

In Eduardo's example of rolling a die, the total number of possible outcomes is simply the number of faces of the die, in this case, 6. The outcome we want is rolling an odd number—1, 3, or 5. To find the probability,

$$P(\text{odd}) = \frac{\text{number of ways an odd number occurs}}{\text{total number of possibilities}} = \frac{3}{6} = 0.5.$$

TEDDY: *So saying the probability is $\frac{1}{2}$ is the same thing as saying there's a 50% chance. That's why you can both be right.*

EDUARDO: *Who is righter?*

TEDDY: *Most of our probabilities will be fractions. It's usually the way mathematicians write them.*

EDUARDO: *OK, I can live with that. And a probability of 0 means the event can't happen.*

SOPHIE: *And a probability of 1 means that it always happens, 100% of the time?*

EDUARDO: *Exactly. And all that work that Sam, Ethan, and Tatiana did in the last chapter gave us the number of possibilities for lots of different events.*

TEDDY: *Well then, let's do a probability problem with SET.*

SOPHIE: *Here's a question we should be able to answer.*

Question 1: What is the probability that three randomly chosen cards form a *SET*?

This is an excellent first question to consider. The event we care about is creating a *SET* with three cards. There are 1080 *SET*s in the deck, so that's going to be our numerator. For our denominator, we need the total number of ways to choose three cards.

TEDDY: *I know how to do that! It has to be the combination 81 choose 3 because we want the total number of ways to choose any three cards!*

Teddy's right. The probability that three randomly chosen cards form a *SET* is

$$P(SET) = \frac{\text{number of } SET\text{s}}{\text{number of three-card subsets}}$$

$$= \frac{1080}{\binom{81}{3}} = \frac{1080}{85{,}320} = \frac{1}{79} \approx 1.27\%.$$

SOPHIE: *Cool! But I think there's an easier way to do this problem. What if I pick just two cards at random from the deck. How could I complete a SET?*

EDUARDO: *Well, there's only one card out of the remaining 79 cards that completes the SET.*

SOPHIE: *Right—the fundamental theorem of SET in action! And, since each of the 79 cards is equally likely to be chosen as the third card, the answer is just $\frac{1}{79}$.*

EDUARDO: *That's a nice, quick solution. But either way, there's only a 1.27% chance of picking a SET at random. That's tiny! It means that grabbing three cards at random from the layout is not a good strategy for finding SETs.[2]*

SOPHIE: *It makes sense though, since there are so many ways to choose three cards at random. But here's another question. When you start the game, you lay out 12 cards. What is the probability that there are no SETs in the initial layout?*

Question 2: Choose 12 cards at random from the deck. What is the probability that there are no *SET*s among these cards?

This is one of the most common questions to ask when working on probability problems in SET. It's also important because the whole

---

[2] Well, duh.

point of the game is to find *SET*s in layouts of 12 cards. Unfortunately for Sophie (and the rest of us), the calculation of this probability is so difficult that no one has yet calculated it exactly.[3]

That doesn't mean we should give up, though. One of the best strategies for attacking a hard problem is to first solve a simpler version of the problem. Here, we can change 12 to a smaller number. If we start with just three cards (instead of 12), then we're already done: Teddy just showed that the probability that three random cards are a *SET* is $\frac{1}{79}$, so the probability that three randomly chosen cards *do not* form a *SET* is $1 - \frac{78}{79} = 0.9873$.

TEDDY: *So* P(*SET*) $= \frac{1}{79}$ *and* P(no *SET*) $= \frac{78}{79}$. *We're* 100% *sure that either the three cards will be a SET or they won't since* $\frac{1}{79} + \frac{78}{79} = 1$.
EDUARDO: *Nicely done, Einstein.*
SOPHIE: *I think we should try more cards. How about four? If we can figure out the probability of having a SET in any four-card configuration, then we can figure out the probability of not having a SET in a layout of four cards.*

Question 3: Choose four cards at random from the deck. What is the probability that there are no *SET*s among these cards?

We'll use Sophie's idea to first find the probability that the four cards do contain a *SET*, then subtract our answer from 1. For the denominator, there are $\binom{81}{4} = 1,663,740$ ways to pick four cards from the deck, so that will be the total number of possibilities.

For the numerator, we need to figure out how many ways to lay out four cards that *do* contain a *SET*. To do this, first pick one of the 1080 *SET*s, and then pick one more card. Once a *SET* is chosen, there are 78 cards left in the deck, making the total number of layouts of four cards with a *SET* $1080 \times 78 = 84,240$:

$$P(SET) = \frac{84,240}{1,663,740} = \frac{4}{79} \approx 5.06\%.$$

---

[3] Or, if someone has, they haven't told us about it.

EDUARDO: *That fraction really simplified nicely! It's exactly four times as big as the probability we found for three cards. Is that a coincidence?*

SOPHIE: *I don't think so. Let's say our four cards are A, B, C, and D. Then there are four possible SETs: ABC, ABD, ACD, or BCD.*

TEDDY: *But we know the probability that any of those form a SET: it's $\frac{1}{79}$.*

EDUARDO: *I see! Since each of these four possibilities has a probability of $\frac{1}{79}$, and there's no overlap, we can get our answer to the four-card question by adding $\frac{1}{79}$ four times.*

TEDDY: *That's cool—we found a simpler way to do the problem after we saw the answer!*

EDUARDO: *And there's no overlap because if ABC is a SET, then it's impossible for any of the other combinations to be a SET.*

Returning to question 3, the probability of having a layout of four cards that do not contain a SET is

$$1 - \frac{4}{79} = \frac{75}{79} \approx 94.9\%.$$

A third way to find this probability appears in exercise 3.1.

SOPHIE: *That was pleasant. What if there are five cards at the beginning?*

TEDDY: *Well, if the probability for four cards is 4 times the probability for three cards, I wonder if we can just multiply by 4 again? Or maybe there's another pattern?*

EDUARDO: *In general, it's a good idea to look for patterns. But sometimes they don't work.*

This time, we'll see that Eduardo is right.

Question 4: Choose five cards at random from the deck. What is the probability that there are no SETs in the five cards?

When we try to extend the count to a five-card layout, things get more complicated. The reason is that there are now two possibilities for

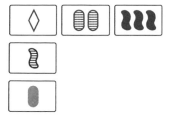

Figure 3.1. These five cards contain a pair of intersecting *SET*s.

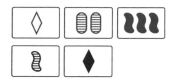

Figure 3.2. These five cards contain exactly one *SET*.

finding a *SET* within five cards. We could have a pair of intersecting *SET*s, in which case we have a layout of five cards containing two *SET*s as in figure 3.1, or we could have only a single *SET*, as in figure 3.2.

We can answer question 4 completely by adding the probabilities for these two non-overlapping cases. (When two events don't overlap, then the probability that both events happen is the sum of the two individual probabilities. People say such events are *disjoint* or *mutually exclusive*.)

Case 1. There are two *SET*s among the five cards. This case is shown in figure 3.1, and the count is fairly easy. We begin by picking a card. As usual we have 81 choices. Then we want to build two *SET*s around that card. Because each card lives in 40 different *SET*s, there are $\binom{40}{2}$ ways to do this. (We used the same idea to count the number of intersets with a given center in the last chapter.) This gives us the probability for this case:

$$P(\text{two } SET\text{s in five cards}) = \frac{81 \times \binom{40}{2}}{\binom{81}{5}} = \frac{63,180}{25,621,596} = \frac{15}{6083} \approx 0.25\%.$$

Case 2. There is exactly one *SET* among the five cards. This arrangement is shown in figure 3.2, and the counting is a little more complicated. First, we pick a *SET* out of the 1080 *SET*s available. We now have 78 cards remaining, and we need to pick two of them.

There are $\binom{78}{2}$ ways to do this, but we must be careful: the two cards we pick cannot form a *SET* with one of the three cards in our *SET*. Call such a pair of cards a *bad pair*.

How many bad pairs are there? Looking at the *SET* in figure 3.2, note that each of those three cards is contained in an *additional* 39 *SET*s. That means that there are 39×3 bad pairs. Then the number of *good* pairs is just $\binom{78}{2} - 39 \times 3 = 2886$. So we can complete this case:

$$P(\text{one } SET \text{ in five cards}) = \frac{1080 \times 2886}{\binom{81}{5}}$$

$$= \frac{3,116,880}{25,621,596} = \frac{740}{6083} \approx 12.17\%.$$

If we add these two probabilities, we find that the probability of finding *at least one SET* in an arrangement of five cards is $\frac{755}{6083} \approx 12.41\%$.

Putting all the pieces together finally lets us calculate the probability that there is *no SET* in a random arrangement of five cards, completely answering question 4:

- The probability of having a layout of five cards that do not contain a *SET* is

$$\frac{5328}{6083} \approx 87.59\%.$$

TEDDY: *Nuts. I just checked: the probability of no SETs in five cards is not a nice multiple of the probability of no SET in three cards or in four cards. It really does look like there's no nice pattern.*

SOPHIE: *I guess my question was a lot harder than I thought. That stinks.*

EDUARDO: *I imagine with each additional card, it gets even more complicated.[4] No wonder no one has figured it out yet.*

SOPHIE: *At least we know some stuff. We know that the probability of having no SETs in a layout of three cards is about 99%, with four cards it's about 95%, and with five cards it drops to around 88%.*

---

[4] Eduardo has a good imagination. But you can do the six-card problem in exercise 3.2.

EDUARDO: *That's good news—the probability of having no SETs should go down each time we add a card.*

TEDDY: *Right, because each time we add another card to the layout, there are so many more possibilities for finding SETs.*

While we can't determine the exact probability of having no *SETs* in a layout of 12 cards, we can estimate it using computer simulations. These simulations tell us that the probability that there are no *SETs* in an initial layout of 12 cards is around 3.2%. Put another way, this means that 12 cards *do* contain a *SET* nearly 97% of the time.

***Takeaway Message*:** If you're playing the game and struggling to find a *SET* in the initial layout, keep looking![5]

It turns out that even having 15 cards doesn't guarantee that there will be a *SET*. This has happened to us in playing the game, but it's quite rare. In fact, there are collections of 20 cards that have no *SETs*; the guarantee doesn't come until we reach 21 cards.

The proof that there must be a *SET* among any collection of 21 cards uses some beautiful geometry. We'll discuss this more in chapter 5 and again in chapter 9. There are a lot of other probabilities we can't compute by hand, though, and those will be explored in more detail in chapter 10.

3.3 EXPECTED VALUE

Sophie, Eduardo, and Teddy lay 12 cards on the table, as though they are starting a game. The cards they put down are shown in figure 3.3. Instead of removing *SETs*, they try to find all the *SETs* in the layout. (How many did they find?)

TEDDY: *What's on your mind now?*

EDUARDO: *Well, we found three SETs in this 12-card layout (in figure 3.3). I wonder how many SETs there are in a typical layout of 12 cards.*

SOPHIE: *I think that's what expected value is all about!*

---

[5] 97% of the time.

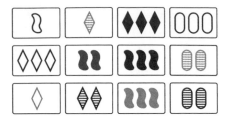

Figure 3.3. Opening layout of a game. How many *SET*s?

Intuitively, expected value measures what happens in the long run.[6] When we analyze data sets, we usually use the word "average" or "mean," but when we discuss probability, we refer instead to *expected value*.

Here's a standard example: What is the expected value of the roll of a die? If the die is fair, we assume that each number is equally likely to appear. In six rolls, we "expect" each number to come up once. So the average of these six rolls is

$$\frac{1+2+3+4+5+6}{6} = 3.5.$$

Alternatively, the probability the die will come up 1 is $P(1) = \frac{1}{6}$, and the probability it comes up 2 is $P(2) = \frac{1}{6}$, and so on. Then the *expected value* is

$$(P(1) \times 1) + (P(2) \times 2) + \cdots + (P(6) \times 6)$$
$$= \left(\frac{1}{6} \times 1\right) + \left(\frac{1}{6} \times 2\right) + \cdots + \left(\frac{1}{6} \times 6\right) = 3.5.$$

Of course, this tells us nothing about what will happen on any given roll. But if we roll a die 100 times and keep track of the sum of the rolls, we expect the sum to be approximately $100 \times 3.5 = 350$.

Back to SET. A very natural question to ask is the following:

Question 5: What is the expected value of the number of *SET*s in the first layout of 12 cards?

---

[6] "In the long run, we are all dead." (John Maynard Keynes)

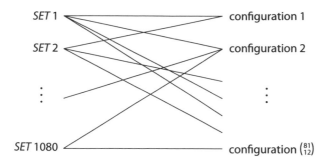

Figure 3.4. Incidence count between *SET*s and 12-card layouts.

Fortunately, it's quite straightforward to answer this question. This stands in sharp contrast to the probability question that stumped us in the last section, namely the probability that the initial 12 cards contain no *SET*s.

To answer this question, we will use an incidence count. We met incidence counts in chapter 2. This time, we list all of the 1080 *SET*s on the left side. On the right, we list all possible layouts of 12 cards. We know (from our work in the last chapter) that there are $\binom{81}{12}$ such layouts. Each *SET* on the left side will be connected to each 12-card layout on the right side that contains that *SET*, as shown in figure 3.4.

The expected value is the *average number of SETs* per 12-card layout. Here's the probability connection. Each subset of 12 cards on the right-hand side is equally likely. Concentrating on the right-hand side of the diagram, the expected value for the number of *SET*s per configuration will be the total number of lines in this diagram divided by $\binom{81}{12}$. So we need to find the total number of lines in the diagram.

As with any incidence count, the total number of lines is the same when counting from the left as when counting from the right. Let's concentrate on the left side.

We determine the number of lines leaving the left side as follows. First, pick a *SET*, and then build a 12-card configuration around it. Our *SET* uses 3 of the 12 cards, which leaves 9 cards left to complete the configuration and 78 cards to choose from. This shows that each *SET* lives in $\binom{78}{9}$ 12-card configurations, so each *SET* on the left side will have $\binom{78}{9}$ lines connecting to the right side. This gives us a total of $1080 \times \binom{78}{9}$ lines.

Since we now know the total number of lines in the configuration, we can solve for the expected value:

$$\text{expected value} = \frac{1080 \times \binom{78}{9}}{\binom{81}{12}} \approx 2.78.$$

This means we can expect on average between 2 and 3 *SET*s to appear in our initial configuration.

There is something beautiful about this calculation. We needed to count the total number of lines in the diagram, which is easy if we look at the left side, but very difficult if we concentrate on the right side. Using the count from the left side gave us our expected value directly. But if we concentrate solely on the right side, letting P(0) represent the probability that there are no *SET*s in the layout of 12 cards, P(1) the probability there is exactly one *SET* in the layout, and so on, we should have $EV = P(0) \times 0 + P(1) \times 1 + P(2) \times 2 + \cdots$.

But we *don't know* P(0) or P(1) or P(2) or anything about these probabilities! (In fact, our inability to compute P(0) is what stopped our calculations in the last section.) The problem is that, unlike our previous incidence counts, not every configuration will have the same number of *SET*s. So, while each *SET* is in the same number of configurations (the left side of the diagram is *regular*), the configurations behave differently: some configurations will have no lines (the configurations containing no *SET*s), and some will have many. (The maximum number of *SET*s in a 12-card configuration is 14. Can you figure out how? You'll see this in project 5.1.)

SOPHIE: *I like expected value; it's really useful. The answer of 2.78 SETs makes sense—we usually find two or three SETs in our initial layouts.*

EDUARDO: *Yeah, even though we couldn't figure out the exact probabilities for our questions about not finding SETs, it's good to understand the averages.*

TEDDY: *Remember how we had easy ways to do some of the probability questions in the last section? I'll bet we could find another way to get this expected value.*

SOPHIE: *Well, there are $\binom{12}{3}$ ways to pick 3 cards from the layout, and the probability that any one of them is a SET is $\frac{1}{79}$. Shouldn't this tell us the expected number of SETs is just $\binom{12}{3} \times \frac{1}{79}$?*

EDUARDO gets a calculator: *It's the same!! I get 2.78.*
TEDDY: *That seems like magic!*

This clever idea works in general—it uses the *linearity of expected value*. It feels like magic because what happens in one event may affect others—for example, if the first three cards form a *SET*, then we know cards 1, 2, and 4 cannot form a *SET*. But that doesn't matter for an expected value calculation. We'll use this approach later in this chapter, and we'll return to it in chapter 7.

We can also find the expected number of *SET*s in an initial configuration of 9 or 15 cards. Using either more incidence counts or Sophie's cleverness, we find the following: in a random collection of 9 cards, we expect approximately 1.06 *SET*s, and the calculation for 15 cards gives us about 5.76 *SET*s.

SOPHIE: *The expected value also shows why 12 is the right number of cards to start the game with. For 9 cards, there aren't enough SETs, on average, and there are too many SETs (and too much scanning through all those cards to find them) with 15 cards. I really like expected value!*
TEDDY: *You already told us. Unfortunately, neither of these ideas works to find the expected number of SETs in the* second *layout of 12 cards, and layouts later in the game.*
EDUARDO: *But I wonder if we could answer similar questions for intersets in initial layouts....*

3.4 FROM *SETS* TO INTERSETS AND PROBABILITY

Let's turn to some questions about intersets.

SOPHIE: *If I choose four cards at random, what are the chances they form an interset?*
TEDDY, sheepishly: *I don't remember what an interset is.*
EDUARDO, patiently: *An interset has four cards. You get one when you take two SETs that contain the same card, then remove that card.*
TEDDY: *Oh yeah, now I remember. What if the card that completes both SETs is also present? Does that matter?*

SOPHIE: *No—I assume that you're talking about searching for intersets in some layout of 12 cards. In that case, all we care about are the four cards. If the card that completes the two SETs is there, too, the four cards are still an interset.*

Question 6: Suppose you choose four cards at random from the SET deck. What is the probability they form an *interset*?

We now have the tools to answer this question very quickly. In chapter 2, we found that there are 63,180 intersets in the deck. Since the number of ways to choose four cards from the deck is $\binom{81}{4}$, we get

$$P(\text{interset}) = \frac{63,180}{\binom{81}{4}} = \frac{3}{79} \approx 3.8\%.$$

TEDDY: *That's interesting! The probability that four random cards form an interset is exactly three times the probability that three random cards form a SET.*

SOPHIE: *OK, so how about the expected value? How many intersets can we expect to find in the initial 12-card layout?*

Question 7: What is the expected value for the number of intersets in an initial layout of 12 cards?

This question has a surprising answer.[7] We first used the clever idea Sophie came up with (linearity of expected value) when we found the expected number of *SET*s in an initial 12-card layout.

Here's how this works. First, there are $\binom{12}{4}$ ways to select four cards from an initial layout of 12 cards. Each four-card subset has a probability of $\frac{3}{79}$ of being an interset (this is the calculation we just did, answering question 6). Then the expected value is the product of these two numbers:

$$EV(\text{number of intersets in 12 cards}) = \frac{3}{79}\binom{12}{4} \approx 18.8.$$

Just to be safe, we'll do this again, using a very quick incidence count. Place the 63,180 intersets on the left and the $\binom{81}{12}$ possible layouts on the right. Since each interset is contained in $\binom{77}{8}$ layouts of 12 cards, there

---

[7] It was surprising to the authors, at least.

are a total of $63,180 \times \binom{77}{8}$ lines in the incidence diagram. Dividing by $\binom{81}{12}$, we get the expected value:

$$\text{EV(number of intersets in 12 cards)} = \frac{63,180 \times \binom{77}{8}}{\binom{81}{12}} \approx 18.8.$$

SOPHIE: *Wait, wait, wait, wait! Wait! That cannot possibly be right! That's huge!*

EDUARDO: *But we definitely did that correctly, twice, and the math doesn't lie. So we better try to understand why there are so many intersets.*

TEDDY: *OK, let's start at the beginning. What makes an interset?*

SOPHIE: *Well, you need two pairs of cards that have the same card that finishes a SET with them.*

EDUARDO: *So we need to look at all the ways we can choose four cards from the layout and then pair them up.*

TEDDY: *Hmm. There are $\binom{12}{4} = 495$ ways to get four cards.*

SOPHIE: *But, for each collection of four cards, we could pair them up three different ways: if the cards are A, B, C, and D, then we could pair them as AB and CD, or AC and BD, or AD and BC.*

EDUARDO: *So that means there are really $495 \times 3 = 1485$ things to check.*

SOPHIE: *Well, for comparison, when we were looking for SETs, we needed to check only $\binom{12}{3} = 220$ subsets. The number of things to check now is almost 7 times bigger!*

TEDDY: *And the expected number is almost 7 times bigger! Spooky! Let's look at our cards and see how many intersets we can spot.*

The group goes back to the 12 cards of figure 3.5; these are the same cards they met in figure 3.3. The goal this time is to find the total number of intersets in the layout. They discover rather quickly that searching for intersets is much slower than searching for SETs.

TEDDY: *You know, we could make a new game out of this! Lay out 12 cards, and start writing down the cards that make intersets. Set a timer, and see who has the most written down in that time.*

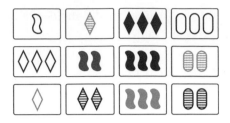

Figure 3.5. Opening layout of a game. How many intersets?

SOPHIE: *That might be a fun variation of* SET. *So, how many did we find, all together?*

In this layout, there are 17 cards that complete lines with two pairs (making 17 intersets), while 1 card completes lines with 3 pairs (making 1 triple interset). All of these are listed at the bottom of the page, and, in this case, none of the cards that complete intersets are present in the layout.[8]

In our expected value for intersets calculation, a triple interset will get counted 3 times, since three separate pairs will make an interset. By the same reasoning, a quadruple interset will get counted $\binom{4}{2} = 6$ times. How many triple intersets should we expect to see in our 12 card layout? How about quadruple intersets? See exercise 3.3.

We know that the largest number of SET-free cards is 20. What about intersets?

Question 8: What's the largest number of cards with no intersets?

A related question has been addressed in the article "Sets, planets, and comets," by M. Baker et al., *College Mathematics Journal*, **44**, no. 4 (September 2013), 258–264. In that paper, the (many) authors defined a *planet* to be four cards that were in the same plane. So a planet consists of four cards that are either an interset, or a SET plus some other card.

---

[8] We shuffled a deck like crazy, cut it, and these were the first 12 cards. We promise. The card that completes the triple interset is 1 Purple Solid Squiggle, while the 17 cards completing ordinary intersets are 3 Green Striped Squiggles, 3 Red Solid Diamonds, 3 Red Solid Squiggles, 3 Red Striped Diamonds, 3 Purple Solid Ovals, 3 Purple Striped Ovals, 3 Purple Striped Diamonds, 2 Red Empty Diamonds, 2 Red Striped Squiggles, 2 Green Empty Ovals, 2 Red Striped Ovals, 2 Purple Empty Diamonds, 2 Purple Striped Squiggles, 1 Red Empty Diamond, 1 Red Empty Oval, 1 Red Solid Oval, and 1 Purple Empty Diamond.

Figure 3.6. Sara's card and Eli's card.

In their paper, the authors did a computer search to show that 10 cards must contain a planet, but there are layouts of 9 cards that don't contain a planet.

One consequence of their work is that the answer to question 8 is at least 10. However, since planets are not the same as intersets, they did not show that every collection of 10 cards contains an interset. Clearly, there's work left to be done.

3.5 FINAL QUERIES

This last section explores some fun probability and expected value problems that come from choosing two cards at random from the deck and exploring the number of different attributes.

Question 9: Two friends, Sara and Eli, pick two cards at random from the deck. How many attributes will they share?

For example, suppose Sara and Eli each pick a card from the deck. What are the chances that their two cards will share no attributes? One attribute? Two or three attributes? What happens on average—what is the expected number of attributes their two cards will share?

To make this concrete, suppose Sara and Eli choose the two cards in figure 3.6.

These two cards differ in three attributes: number, shading, and shape.

Why would we care about this question? Well, once we've picked two cards, we know that there's only one way to complete a *SET* with those two cards (by the fundamental theorem of SET!). Moreover, the number of different attributes in that *SET* will match the number of different attributes of our two cards. So we can model all of our questions about the number of different kinds of *SET*s in terms of selecting two cards from the deck.

Figure 3.7. The *SET* containing Sara's and Eli's cards.

Returning to our example, we form the *SET* containing the cards Sara and Eli chose (figure 3.7). This *SET* has one attribute the same (color) and the other three attributes different.

Let's compute the probability that our two cards are different in all four attributes. Suppose Sara picks 1 Red Empty Diamond, as above. How many cards can Eli pick that are different in every attribute?

First, number: Sara's card has 1 symbol, so Eli's card must have 2 or 3 symbols. For color, Sara's card is red, so Eli's must be green or purple. Similarly, Eli has two choices for the shape and two for the shading. This gives $2 \times 2 \times 2 \times 2 = 16$ choices for Eli's card. Since Eli has 80 cards to choose from, but only 16 match our requirements, we see that the probability their two cards are different in all attributes is $\frac{16}{80} = 20\%$.

We can perform these calculations when the two cards differ in one, two, or three attributes, too. If two cards are randomly chosen from the deck, the probability that they differ in

- all four attributes is 20%;
- exactly three attributes is 40%;
- exactly two attributes is 30%;
- exactly one attribute is 10%.

Here's the connection with *SET*s: Back in chapter 2, we found that exactly 20% of the *SET*s in the deck have all four attributes different, 40% have three attributes different, 30% have two attributes different, and 10% of the *SET*s in the deck have one attribute different. These answers match perfectly! In chapter 7, we'll explain how this connection can be exploited to get excellent approximations when we consider (much larger) decks with (many) more attributes.

SOPHIE: *That was fun—I feel like I understand the game better now.*

EDUARDO: *Yeah, we answered some interesting probability questions and we saw a few different ways to compute expected value.*

SOPHIE: *Expected value is great—it gives you information about how many SETs and how many intersets you should see when you play the game. Did I tell you that I really like expected value?*

EDUARDO AND TEDDY: *It may have slipped out once or twice.*

SOPHIE: *I know what I want to do now—isn't it time for a game?*

EDUARDO AND TEDDY: *YES!*

EXERCISES

EXERCISE 3.1. Question 3 in this chapter asked for the probability that there are no *SET*s among four randomly chosen cards. We solved this problem in two different ways in the chapter; the answer is $\frac{4}{79}$. Here's a third approach.

a. First, choose three cards at random, A, B, and C. What is the probability that your three cards do not form a *SET*?
b. Now choose a fourth card, D. Find the probability that ABD, ACD, and BCD are not *SET*s by figuring out how many choices there are for D.
c. Multiply your answers from parts (a) and (b) to get the probability that there are no *SET*s among the four cards. Explain why this works.

EXERCISE 3.2. Questions 2, 3, and 4 of this chapter found the probabilities that there are no *SET*s in three, four, or five randomly chosen cards. Figure out the probability that there are no *SET*s in six cards. [A possible first step is given below. Feel free to ignore it.]

- You can have either one, two, or three *SET*s among six cards. Count the number of possibilities for each of these three situations.

EXERCISE 3.3. A *triple interset* is a collection of six cards that can be grouped into three pairs, where one card completes a *SET* with each pair. For example, the six cards shown in figure 3.8 form a triple interset that is completed by the 2 Red Solid Diamonds card.

a. Show that if six cards can be paired to form a triple interset, then the pairing is unique, i.e., there is only one way to pair up three cards to do this.

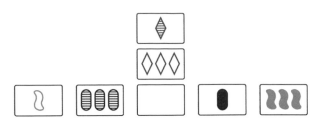

Figure 3.8. Exercise 3.3.

b. Count the number of triple intersets in the deck. [Hint: Modify the argument used to count the number of intersets.] Then use your count to compute the probability that six randomly chosen cards form a triple interset.

c. Use your answer to part (b) and an incidence count to compute the expected number of triple intersets in an initial layout of 12 cards. Then check your answer by looking for triple intersets in a few layouts.

d. (*if you're adventurous, or a little loopy*). Repeat parts (b) and (c) for eight-card *quadruple intersets*.

EXERCISE 3.4. Suppose that two friends, Sara and Eli, each pick a card from the deck. Find the following probabilities. [Hint: Your answers should match the answers given near the end of section 3.5.]

a. Calculate the probability that two random cards differ in exactly three attributes.

b. Calculate the probability that two random cards differ in exactly two attributes.

c. Calculate the probability that two random cards differ in exactly one attribute.

PROJECTS

PROJECT 3.1. This project is devoted to what happens at the end of the game when there are six cards left (figure 3.9). First, break up the cards into three pairs. For example, you might pair them as in figure 3.10.

Now, for each pair, find the card that completes the *SET*. In this example, we get the three cards in figure 3.11.

Notice that these three cards form a *SET*! We'll prove this in chapter 4, but you should try to prove it yourself now. The *SET* formed has two attributes the same (color and shading) and two that differ (number and shape).

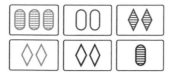

Figure 3.9. Six cards at the end of a game.

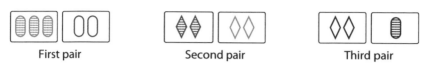

First pair   Second pair   Third pair

Figure 3.10. The pairings.

Figure 3.11. These three cards complete the three *SET*s.

What happens if we pair the cards in a different way? We'll always get a *SET*, but it's possible that the "*SET*" we get will consist of the same card, repeated three times.

Here's your job for this project:

a. First, for the six cards given in figure 3.9, break up the cards into three pairs in every possible way. [Hint: There are 15 ways to break up the cards into three pairs.]

b. For each pairing, complete the *SET*s to get three new cards.

c. For each of the new *SET*s created, determine the number of attributes that are different.

d. The six cards have now produced 15 different *SET*s. How many times did you get all attributes different? How many times were three attributes different and one the same? Let x_0 be the number of *SET*s produced with zero attributes the same, x_1 the number with one attribute the same, etc. (Note that $x_4 = 1$ precisely when the cards form a triple interset.) Call the *signature* of the six cards the list of these values: $\{x_0, x_1, x_2, x_3, x_4\}$.

e. (This is the main part of this project.) There are $\binom{19}{4} = 3876$ possible signatures (this is the number of subsets, with repetition allowed, of size 15 from the *SET* $\{0, 1, 2, 3, 4\}$). How many of these potential signatures actually occur among six-card configurations that arise at the end of a game?

SET and Modular Arithmetic

4.1 WHAT IS MODULAR ARITHMETIC?

Three more of your friends, Shamella, Erin, and Tyler, heard what everyone else was up to and decided to join the fun and become characters in our book. They have just begun college, and Shamella is eager to tell them about her new math course, number theory.

> SHAMELLA: *Today my math teacher showed us some interesting applications of modular arithmetic.*
> ERIN: *What is modular arithmetic?*
> TYLER: *Wait, we saw this in the first chapter! It's sometimes called clock arithmetic.*
> ERIN: *Oh yeah, I remember now! We did a clock problem. We learned how to figure out what time it will be 100 hours from now. Or any number of hours from now.*
> SHAMELLA: *Yeah, clocks are the most natural example of modular arithmetic. Days of the week can also work, and months, because calendars are cyclic ... and also music scales, because those repeat! But my teacher showed us some uses that answer more abstract questions.*

Before our group continues, here's a reminder of the notation we mentioned in chapter 1: we will write $a = b$ (mod c) if a and b have the same remainder when divided by c. So, as we saw in chapter 1, $110 = 14$ (mod 24), because when you divide 110 by 24, you get a remainder of 14. (One quick note: "mod" is short for "modulo." We also say 24 is the "modulus" we're working with here.)

4.2 MODULAR ARITHMETIC PROBLEMS

Our three students find the following problem in Shamella's homework assignment.

Question 1: What is the last digit of 1987^{1987}?

ERIN: *I have no idea how to do this problem.*

TYLER: *Me neither! I just typed it into my calculator, and it said "OVERFLOW."*

ERIN: *Great—I'll just write that down for the answer. Thanks.*

SHAMELLA: *All right everyone, calm down. This isn't as hard as it looks. We can start by doing simpler versions of this problem.*

ERIN: *Okay, well, $1987^1 = 1987$. Done!*

TYLER: *What about 1987^2? That's too big for me to do in my head, but we only need the last digit of 1987^2. Maybe we could just square the last digit of 1987. If we square 7 we get 49, so I'm guessing the last digit of 1987^2 is going to be 9 too. I can check this on my calculator—$1987^2 = 3,948,169$, so I was right. Cool!*

SHAMELLA: *Yeah, that always works—when we're asked for the last digit of some number raised to a power, the last digit of the number we're given is all we need to worry about.*

ERIN: *So we just need to do 7^{1987}! But that's still too big, isn't it?*

TYLER: *It certainly is. But maybe we can find a pattern. For 1987^n, the last digit is 7 when $n = 1$, and 9 when $n = 2$. What about when $n = 3$?*

SHAMELLA: *Yes, this is how to do the problem. For $n = 3$, we need to do 7^3, which is 343.*

ERIN: *Couldn't we also have just said $7^2 = 49$, and then $7 \times 9 = 63$, so the last digit has to be 3?*

TYLER: *Yes! This is much easier than I thought. Okay, for $n = 4$, I just have to multiply 3, which was our previous last digit, by 7. When I multiply 3 by 7 I get 21, so the last digit is 1. And then for $n = 5$, we'll be back to 7, because $1 \times 7 = 7$.*

SHAMELLA: *Sweet, we found a pattern! After 7 we have to go back to 9 because $7 \times 7 = 49$ and so on. That's always what* *happens with these problems—the last digit will cycle through some pattern and then repeat. So our pattern goes 7, 9, 3, 1, 7, 9, 3, 1, ...*

ERIN: *Oh I see. So the last digit of 1987^n will always be either 7, 9, 3, or 1, and the digits will always cycle through in that order.*

TYLER: *This must be where modular arithmetic comes in! Our cycle is four numbers, so we should work mod 4. For 1987^n, when $n = 1$ (mod 4), the last digit is 7; when $n = 2$ (mod 4), the last digit is 9; when $n = 3$ (mod 4), the last digit is 3; and when $n = 4$ (mod 4), the last digit is 1. So now we just need to figure out what 1987 is mod 4.*

SHAMELLA: *Yup! And that should be easy enough. One note though: 4 (mod 4) = 0 (mod 4).*

TYLER: *Oh, that makes sense. Anything that is a multiple of 4 is 0 (mod 4).*

SHAMELLA: *Yeah, whenever we work mod n, $n = 0$ (mod n). The biggest we can ever get mod n is $n - 1$.*

ERIN: *That means anything that is a multiple of n is going to be equivalent to 0 (mod n).*

TYLER: *So we really just need to figure out how far away 1987 is from a multiple of 4.*

ERIN: *Can't we just divide 1987 by 4 to see what the remainder is?*

SHAMELLA: *Yes! Okay, my calculator says $1987 \div 4 = 496.75$. Where does that get us?*

TYLER: *Well 0.75 is $\frac{3}{4}$, so I guess the remainder is 3. But just to verify this really fast on my calculator, $496 \times 4 = 1984$, so then $1987 = (496 \times 4) + 3$.*

SHAMELLA: *Great! So $1987 = 3$ (mod 4). Now we just have to remember what the last digit was when $n = 3$. It looks like it was 3.*

ERIN: *So the last digit of 1987^{1987} is 3.*

TYLER: *Ta da!*

SHAMELLA: *... That was sort of anticlimactic. Let's try another problem!*

TABLE 4.1.
The possible values of a^2, b^2, and c^2, (mod 3).

a^2	b^2	$a^2 + b^2 = c^2$	Possible?
0	0	0	yes
0	1	1	yes
1	0	1	yes
1	1	2	no

The group gathers around the number theory book, and finds the next question:

Question 2: Suppose $a^2 + b^2 = c^2$ for positive integers a, b, and c. Show that either a or b (or both) must be a multiple of 3.

SHAMELLA: *This will be more climactic. This is a proof! Plus, it's about the Pythagorean theorem, which is cool.*

ERIN: *Okay, but once again, I have no idea where to start.*

TYLER: *Once again, me neither. I know we're going to be working mod 3, but that's all I know.*

SHAMELLA: *Well, it involves squaring each of a, b, and c, so let's start by squaring numbers, mod 3.*

ERIN: *Okay, so, if I have a number equivalent to 0 (mod 3), and I square it, $0^2 = 0$, so the square will still be 0 (mod 3).*

TYLER: *I guess what you're saying is that the square of a multiple of 3 is still a multiple of 3. We certainly could have figured that out without the mods, though.*

ERIN: *Let's look at non-multiples of 3. If I have a number equivalent to 1 (mod 3), and I square it, $1^2 = 1$, so the square will still be 1 (mod 3). If I have a number equivalent to 2 (mod 3), and I square it, $2^2 = 4$, and 4 = 1 (mod 3), so the square will be equivalent to 1 (mod 3).*

SHAMELLA: *So you just figured out that perfect square numbers can never be equivalent to 2 (mod 3).*

The group decides to organize the information they've discovered in a chart. See table 4.1.

ERIN: *So if we look at the first three rows in your table, then either $a^2 = 0$ (mod 3) or $b^2 = 0$ (mod 3), or both. That means at least one of a or b is a multiple of 3, which is what we were trying to prove!*

SHAMELLA: *But we're not done. Look what happens when $a^2 = 1$ (mod 3) and $b^2 = 1$ (mod 3). In this case, neither one of them is a multiple of 3. But then when I add them, I get that $c^2 = 2$ (mod 3), but c^2 can't be equivalent to 2 (mod 3) since it is a perfect square. So this case is impossible! Now we're really done!*

TYLER: *I agree. In all of the cases that can occur, we had at least one of a or b as a multiple of 3. So there's our proof!*

ERIN: *Cool! It's interesting how we ended up also proving that perfect squares must be equivalent to either 0 or 1 (mod 3).*

SHAMELLA: *I really like number theory. It turns out perfect squares have similar restrictions for other moduli, which is pretty cool.*

TYLER: *"Moduli"?*

ERIN: *That must be the plural of modulus.*

SHAMELLA: *Indeed. When you study number theory, you get to use cool words like "moduli."*

4.3 THAT'S ALL WELL AND GOOD, BUT WHAT ABOUT SET?

Application 1: Three cards are a *SET* if and only if their coordinates sum to $(0, 0, 0, 0)$ (mod 3).

Recall from chapter 1 that we can set up a table to assign numbers to the different attributes. These assignments are arbitrary; we will stick to the assignment in table 4.2 throughout the book.

Our mantra here is going to be Number, Color, Shading, Shape. Start repeating this in your head: Number, Color, Shading, Shape. Number, Color, Shading, Shape. It's useful to keep this in mind when reading the coordinates. (This will also come in handy later, when we learn the End Game.) Now let's pick a *SET*; see figure 4.1.

TABLE 4.2.
Assignment of coordinates to cards.

Attribute	Value		Coordinate
Number	3, 1, 2	↔	0, 1, 2
Color	green, purple, red	↔	0, 1, 2
Shading	empty, striped, solid	↔	0, 1, 2
Shape	diamonds, ovals, squiggles	↔	0, 1, 2

Figure 4.1. From left to right, these cards have coordinates (2, 2, 1, 0), (2, 1, 1, 0), and (2, 0, 1, 0).

ERIN: *I remember this from chapter 1. We can label each card as a vector[1] with four coordinates, using our table.*

TYLER: *Right. Then we need to add up all four coordinates, mod 3.*

The group converts the cards to coordinates, like so:

- 2 Red Striped Diamonds ↔ (2, 2, 1, 0),
- 2 Purple Striped Diamonds ↔ (2, 1, 1, 0), and
- 2 Green Striped Diamonds ↔ (2, 0, 1, 0).

Next, they add up the coordinates for the three vectors, working mod 3:

1. First coordinate: $2 + 2 + 2 = 6 = 0$ (mod 3).
2. Second coordinate: $2 + 1 + 0 = 3 = 0$ (mod 3).
3. Third coordinate: $1 + 1 + 1 = 3 = 0$ (mod 3).
4. Fourth coordinate: $0 + 0 + 0 = 0 = 0$ (mod 3).

SHAMELLA: *That's so neat! All three possibilities for how the coordinates can sum appear here, and they all give 0 (mod 3). That makes sense—we're either adding*

---

[1] Vectors will play an important role later. We think of a vector as an ordered list of numbers, mod 3. If you like vectors, then you will love chapter 8.

Figure 4.2. Two lonely cards. Find the third card to complete their *SET* and make them happy.

$0 + 1 + 2$ *or we're adding the same number three times, so we'll always get a multiple of 3, which is equivalent to* 0 *(mod 3).*

TYLER: *One last question: Is it possible to get faked out by sneaky cards? I mean, could we ever have a combination of three cards whose coordinates add to* 0 *(mod 3) but who are not a SET?*

ERIN: *Hmm....*

Tyler's question is worth thinking about. Our plucky students have shown that, if three cards form a *SET*, the coordinates sum to $(0, 0, 0, 0)$ (mod 3). Tyler's question is about the *converse*: if three cards sum to $(0, 0, 0, 0)$ (mod 3), then they make a *SET*.

Here's why this is true. For each coordinate, the only collections of three numbers that sum to 0 are the ones we found:

- If $a + b + c = 0$ (mod 3), then either $a = b = c$, or $a, b,$ and c are the numbers 0, 1, and 2, in some order.

"Proving" this just amounts to checking that nothing else works. For instance, if $a = b = 1$ and $c = 2$, we have $a + b + c = 1$ (mod 3), so this situation can't occur. We conclude there are no sneaky cards: three cards form a *SET* precisely when their sum is $(0, 0, 0, 0)$ (mod 3).

Application 2: Given any two cards, find the third card to create a *SET*.

Given any two cards, the fundamental theorem of SET tells us there is exactly one card that completes the *SET*. For the cards in figure 4.2, it's easy enough to do this without any coordinates, vectors, or modular arithmetic. But it's worth showing how the techniques of this chapter can be applied to solve this problem.

Figure 4.3. The cards made a friend!

Our group starts by getting coordinates into the problem. First, they find the coordinates for the two cards in the figure:

- 2 Purple Solid Squiggles ↔ (2, 1, 2, 2), and
- 2 Green Empty Squiggles ↔ (2, 0, 0, 2).

Now, concentrating on the vectors (instead of the cards), here's the procedure. Call the vector for the missing card C. Since a *SET* sums to (0,0,0,0), we need

$$(2, 1, 2, 2) + (2, 0, 0, 2) + C = (0, 0, 0, 0).$$

SHAMELLA: *Well, I would simplify this first by adding the first two cards. Since* $(2, 1, 2, 2) + (2, 0, 0, 2) = (1, 1, 2, 1)$, *we get*

$$(1, 1, 2, 1) + C = (0, 0, 0, 0).$$

ERIN: *Now it's easy. We just need* $C = (2, 2, 1, 2)$ *to make the sum 0 in each coordinate. So I guess*

$$C = (0, 0, 0, 0) - (1, 1, 2, 1) = (2, 2, 1, 2).$$

TYLER: *But if I subtract, I get*
$$(0, 0, 0, 0) - (1, 1, 2, 1) = (-1, -1, -2, -1). \text{ What gives?}$$
SHAMELLA: *These are the same! That's the beauty of modular arithmetic:*

$$(-1, -1, -2, -1) = (2, 2, 1, 2) \quad (\text{mod } 3).$$

The group checks table 4.2 to turn the vector (2, 2, 1, 2) back into a card. They discover it's 2 Red Striped Squiggles, as shown in figure 4.3, which they knew from playing the game.

TYLER: *Isn't there an easier way to find the vector for the missing card? Instead of all that arithmetic, I could've looked at those first two coordinates, saw they were both 2, and figured the first coordinate of my missing card had to be 2. And for the second coordinate, if I see a 0 and a 1, I know the second coordinate of my missing card has to be a 2. The same idea works for the other two coordinates, so I get (2, 2, 1, 2). Boom!*

ERIN: *Yeah, you're right—that was faster. For our purposes, I suppose that method should work fine for finding a missing card, but I bet it'll also be useful for some of the material that's coming.*

SHAMELLA: *I love foreshadowing!*

Application 2.5: Find the third card of a *SET* a different way.

Modular arithmetic is powerful enough to give us a few different ways of understanding SET. We can also find the third card using an idea from high-school geometry, of all things.[2]

SHAMELLA: *In fact, there's another way we could have done it. We could have added each coordinate for the two given cards, then divided the result by 2. If the two given cards are* $(2, 1, 2, 2)$ *and* $(2, 0, 0, 2)$, *then the third card is*

$$C = \frac{(2, 1, 2, 2) + (2, 0, 0, 2)}{2}.$$

ERIN: *Wait. How do we divide by 2? What does this even mean?*

Here's what Shamella means. First, "dividing by 2" is the same as multiplying by $\frac{1}{2}$. But $\frac{1}{2}$ is not one of the three numbers 0, 1, or 2 that we are allowed to use when working mod 3.

The way we work around this problem is to interpret $\frac{1}{2}$ as the *multiplicative inverse* of 2, i.e., the number that we need to multiply 2 by to get an answer of 1. But we already know that $2 \times 2 = 1 \pmod 3$, so $\frac{1}{2} = 2 \pmod 3$.

---

[2] Did you see this coming? We didn't.

SET and Modular Arithmetic • 81

SHAMELLA: *So dividing by 2 is the same as multiplying by 2 when we work mod 3! That's really cool, and kinda strange.*

ERIN: *Let's see if this works. I get*

$$C = 2\left((2, 1, 2, 2) + (2, 0, 0, 2)\right)$$
$$= 2(1, 1, 2, 1)$$
$$= (2, 2, 1, 2) \quad (\text{mod } 3).$$

TYLER: *It works—that's the same card we found before! But I have another question. If you add each coordinate and divide the result by 2, you're using the formula that finds the midpoint of a segment. Are you saying that the "midpoint" of two cards would be the card that completes the SET?*

ERIN: *That would make things even weirder! Suppose I have three cards A, B, and C that make a SET. If what you're saying about midpoints is true, then A is the midpoint of B and C, and B is the midpoint of A and C, and C is the midpoint of A and B:*

$$A = \frac{B + C}{2}, \qquad B = \frac{A + C}{2}, \qquad C = \frac{A + B}{2}.$$

That doesn't make sense geometrically!

SHAMELLA: *It doesn't make sense if you're thinking of the usual Euclidean geometry you remember from high school, no. But the geometry in these cards is not our usual geometry.[3] Anyway, the reason it works is because dividing by 2 is equivalent to multiplying by 2 when we're working mod 3.*

TYLER: *I see. First we can sum the two coordinates and then* multiply *by 2 to find our midpoint, which is equivalent to dividing by 2 (mod 3).*

SHAMELLA: *Let's try it! Erin already did this in her head, but let's go through it coordinate by coordinate.*

---

[3] Evidently. See chapters 5 and 9 for more geometry.

Our cards had coordinates (2, 1, 2, 2) and (2, 0, 0, 2). We check the midpoint formula for each coordinate, remembering to multiply by 2 instead of dividing by 2, since we're working mod 3:

1. First coordinate: $2 + 2 = 4$ which is 1 (mod 3), and then $1 \times 2 = 2$, which is 2 (mod 3).
2. Second coordinate: $1 + 0 = 1$ which is 1 (mod 3), and then $1 \times 2 = 2$, which is 2 (mod 3).
3. Third coordinate: $2 + 0 = 2$ which is 2 (mod 3), and then $2 \times 2 = 4$, which is 1 (mod 3).
4. Fourth coordinate is identical to the first, because we have two 2s again. So the last coordinate is 2.

SHAMELLA: *It works: we get that the coordinates of the card that completes the SET should be (2, 2, 1, 2).*

TYLER: *I love how we can do the same problem different ways and always get the same answer. It's nice when math makes sense!*

SHAMELLA: *Math is supposed to make sense. That's its job!*

Application 3: The final three cards.

Having successfully applied some modular arithmetic to a few questions, our trio turn their attention to what happens at the end of the game.

SHAMELLA: *Here's a question. What happens when we have three cards left at the end of the game?*

ERIN: *I don't think that's ever happened.*

TYLER: *No, it's impossible! Our deck has 81 cards, and we know we can organize it into 27 SETs.[4] There are probably a lot of ways to do that, but what that means is that the (mod 3) sum of the whole deck should be (0,0,0,0).*

SHAMELLA: *Here's another way to think about this: If you consider color, then 27 of the cards are green, 27 are purple, and 27 are red. So if you sum all of those numbers, you get*

---

[4] If you've ever cleared the deck while playing SET, then you've "organized" the deck into 27 SETs.

$27 \times 0 + 27 \times 1 + 27 \times 2 = 27 \times 3 = 0 \pmod 3$.
And the same will be true of number, shading, and shape.

ERIN: *Sounds good to me. Now, if we're playing a game, then the sum of the pile of SETs that have been removed should also be (0,0,0,0), assuming they're real SETs and people aren't making mistakes.*

TYLER: *I see where this is going. If we get to the end, and we have three cards left, then we know the pile of SETs we've taken away sums to (0,0,0,0), and we know the entire deck sums to (0,0,0,0), so those last three cards also have to sum to (0,0,0,0).*

SHAMELLA: *So if you get to the end and you have three cards left, those last three cards have to be a SET!*

ERIN: *Yes! It would actually be pretty exciting to have this happen in a game: If you get to the end and there are six cards left, and you find a SET in those six cards, you know the last three are also a SET without having to check. When this happens, you can just immediately yell "SET! SET!" and grab all six cards!*

TYLER: *Well, I don't know that you have to yell. But yes, this should work, as long as you trust the people you're playing with....*[5]

SHAMELLA: *Let's take a break and play until we clear the deck.*

ERIN: *Didn't you read chapter 1? We might have to play almost a hundred games before that happens.*

TYLER: *Or, we can play a variation of the game*[6] *where we get to redistribute SETs at the end, until we are able to clear the remaining cards.*

ERIN: *That'll still probably take a while.*

TYLER: *Then let's get started. We're wasting time talking about it. (Hours pass....)*

---

[5] Let's hope they are playing with a full deck.
[6] This variation is called Clear the Deck and is described in the interlude.

Figure 4.4. Five cards at the end of a game; the sixth is hidden.

4.4 THE END GAME

Application 4: The End Game.

Recall from chapter 1 that we can put aside a card at the beginning of a game, play the game, and then use the remaining cards to determine the identity of the missing card. There's an example in figure 4.4 where six cards are left at the end of a game (played by Shamella, Erin, and Tyler), and we've hidden one of the cards.

SHAMELLA: *How can we use modular arithmetic to find the missing card?*

TYLER: *Remember, the whole deck sums to (0,0,0,0). If the whole deck sums to (0,0,0,0), and each of the SETs that we've removed sums to (0,0,0,0), then the leftover cards should also sum to (0,0,0,0).*[7]

ERIN: *Does that mean that the leftover cards are some sort of super SET? I mean, didn't we decide earlier that cards sum to (0,0,0,0) if and only if they are a SET?*

SHAMELLA: *We did, but that's only true for three cards. But it's actually an interesting question to ask what sort of structure a collection of cards greater than three would have that sum to (0,0,0,0).*[8]

TYLER: *I think the important part is that each individual attribute has to sum to 0. So, if we isolate attributes, we could determine the missing card.*

ERIN: *I guess we could figure out the vector coordinates for each of those five cards, sum them together mod 3, subtract that sum from (0, 0, 0, 0) (mod 3), and that vector would then be the missing card.*

---

[7] "Nothing from nothing leaves nothing." (Billy Preston)

[8] We saw this in chapter 3. If it's nine cards, then it's been called a "comet."

SHAMELLA: *Well, I'm not doing that. There's gotta be a faster way to do this.*

TYLER: *Yeah, I doubt many people would want to do that during an actual game. I sure wouldn't.*

Application 4.5: Different ways of finding the missing card.

There are a few ways to determine the missing card that use ideas from modular arithmetic. We describe two ways.

1. Sort the cards on the table into "single attribute *SET*s," meaning groups of three cards that are all the same or all different in one specific attribute. Here's how to apply this idea:

   - Number: Looking at the cards in figure 4.4, we first find the number for the missing card. Concentrating solely on number (and temporarily ignoring color, shading, and shape), we can put aside a *SET* of 3s. The two remaining cards include one 3 and one 2, so the missing card must have 1 symbol. This makes two "number *SET*s."
   - Color: Looking at color, the first three cards are a "color *SET*." The last two cards are green and red, so the missing card must be purple.
   - Shading: For shading, the middle three cards are a "shading *SET*," and the two on the ends are both empty, so the missing card must be empty.
   - Shape: And finally, looking at shape, the three cards on the right are a "shape *SET*," and the remaining two are a diamond and an oval, so the missing card must have squiggles.

   Putting all of this together gives us 1 Purple Empty Squiggle.[9]
   A last comment about this method: it doesn't matter how you make the attribute *SET*s. Suppose your five cards were purple,

---

[9] How do you keep all of this in your head? Honestly, there are plenty of times we've done this and had something fall out of our brains. Advice: Remember that mantra. Number, Color, Shading, Shape! Repeating this in your head will really help you keep track of the attributes you've already found.

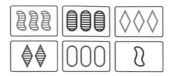

Figure 4.5. Six cards at the end of a game.

purple, purple, green, and red. If you make an all-purple *SET*, then the cards left are red and green, so the missing card must be purple. On the other hand, if you make a purple–green–red *SET*, then the cards left are both purple, and again, you'd find the missing card is purple. You can convince yourself that you'll always get the correct answer no matter how you split up the attribute *SET*s.

2. There's a second way we can use modular arithmetic to find the identity of the missing card. This procedure also works attribute by attribute. The idea is that numbers need to be the same, mod 3. Here's the procedure, explained via the example from figure 4.4.

- Number: Among the five cards, none have 1 symbol, one has 2 symbols, and four have 3 symbols. Writing this in order, we have 0, 1, 4 as the numbers of cards in each category. We need our missing card to make these three numbers the same, mod 3. That means that the missing card has one symbol, so our ordered list will be (1, 1, 4).
- Color: We have two greens, one purple, and two reds. The missing card must be purple in order to make these three numbers equal, mod 3, so our ordered list is (2, 2, 2).
- Shading: There are two empty cards, three striped cards, and no solid cards. To make these the same, mod 3, we need another striped card. This makes the ordered list (3, 3, 0).
- Shape: There are two diamonds, two ovals, and one squiggle. This forces our missing card to be a squiggle, so the ordered list is again (2, 2, 2).

Putting all of this together gives the same answer as before: 1 Purple Striped Squiggle. See figure 4.5.

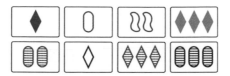

Figure 4.6. End Game. Eight cards at the end of a game, with the ninth hidden. Find the missing card, then find the *SET* it makes with two of the cards here.

TYLER: *I get the first procedure, but why does the second way also work? Why do the totals have to be equivalent, mod 3? Is there really no other possibility?*

SHAMELLA: staring at the cards intently. *I think I get it! Let's just consider one attribute (like color) and organize the cards into "single attribute SETs," like we did before. When you have three cards that are all the same for an attribute, that adds 0, 3, or 6—which are all equivalent to 0* (mod 3)—*to the total for the number of cards with that expression of the attribute. And when you have three cards that are all different for that attribute, that adds three 1s to the totals for each expression of that attribute, which will still keep their totals equivalent, mod 3. So when you look at all the "single attribute SETs" together, the total for each expression will have to be the same, mod 3.*

ERIN: *Modular arithmetic to the rescue!*

TYLER: *Okay, so we've seen the numbers of expressions of an attribute could be* (4, 1, 1), (2, 2, 2), *and* (3, 3, 0). *Are there other configurations we could have with six cards left?*

SHAMELLA: *We could have* (6, 0, 0), *a situation where all six leftover cards share an attribute. So for instance, all six cards are purple, or they're all solid, or they're all ovals. That's probably pretty rare.*

ERIN: *I guess those are all the possibilities for six cards.*

Erin is right—the ordered lists must be rearrangements of (4, 1, 1), (3, 3, 0), (2, 2, 2), and (6, 0, 0). In order to test these ideas, the group hides a card, then plays another game. This time, there are eight cards left. See figure 4.6.

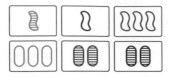

Figure 4.7. Six cards left at the end of a game played by Shamella, Erin, and Tyler.

For a challenge, use one or both of the techniques described above to find the missing card. Then find the *SET* it completes. Answer below.[10]

With nine cards left at the end of a game, some possible attribute-ordered lists are $(3, 3, 3)$, $(5, 2, 2)$, and $(4, 4, 1)$. Are there any more possible lists? See exercise 4.8 for a chance to figure this out on your own.

Finally, you might wonder which technique for playing the End Game is better: the first procedure (mentally removing attribute *SET*s) or the second (making the numbers the same, mod 3). In our experience, the first procedure is faster. You can also use a mixed strategy; find some attributes with one technique and the rest with the other, if you like. But doing so leads to madness.[11]

4.5 THE SIX-CARD THEOREM

Application 5: The six-card theorem.

At this point, our three heroes take a break to play some SET. In the meantime, we'll discuss a theorem, sometimes referred to as the Stupid SET Trick.[12] Suppose six cards are left at the end of the game. If we partition them into pairs, then the cards that complete the *SET*s for each pair will themselves be a *SET*. (See figures 4.7 and 4.8.)

This works no matter how we pair up the cards! Let's try pairing them a different way. (See figures 4.9 and 4.10.)

---

[10] The missing card is 2 Green Solid Ovals, and it forms a *SET* with 1 Red Empty Oval and 3 Purple Striped Ovals.

[11] Again, we speak from experience. It's also slow, and it's very easy to get confused and get the wrong answer for the missing card.

[12] We saw this in exercise 1.2 and project 3.1. It was first proved by one of the authors. Well done, Hannah.

Figure 4.8. The cards that complete vertical *SET*s with the cards from figure 4.7 are themselves a *SET*.

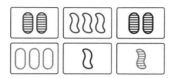

Figure 4.9. The six cards from figure 4.7, rearranged into three different pairs.

Figure 4.10. The cards that complete vertical *SET*s with the rearranged cards from figure 4.9 are also a *SET*.

Why does this work? The explanation relies on modular arithmetic.

We have six cards left at the end of the game. Let's call their corresponding coordinate vectors A, B, C, D, E, and F. We know that $A + B + C + D + E + F = (0, 0, 0, 0)$, because these correspond to cards left at the end of the game.

Now, we'll add three cards, whose coordinates are X, Y, and Z, to complete three *SET*s: our *SET*s are ABX, CDY, and EFZ. This means that

$$A + B + X = (0, 0, 0, 0) \quad (\text{mod } 3),$$

$$C + D + Y = (0, 0, 0, 0) \quad (\text{mod } 3), \quad \text{and}$$

$$E + F + Z = (0, 0, 0, 0) \quad (\text{mod } 3),$$

because these are *SET*s. Thus, adding these three equations gives us

$$A + B + X + C + D + Y + E + F + Z = (0, 0, 0, 0) \quad (\text{mod } 3).$$

Addition is commutative, so we can rearrange those variables:

$$(A + B + C + D + E + F) + (X + Y + Z) = (0, 0, 0, 0) \quad (\text{mod } 3).$$

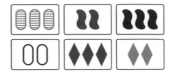

Figure 4.11. Six cards at the end of a game.

But we also know $A + B + C + D + E + F = (0, 0, 0, 0)$ (mod 3). Subtracting equations gives $X + Y + Z = (0, 0, 0, 0)$ (mod 3). This tells us that XYZ is a *SET*. QED.[13]

Application 6: The six-card special case.

While you were reading that proof, Shamella, Erin, and Tyler played a lightning fast round, and they now have six cards left, as shown in figure 4.11.

As above, they want to complete the three vertical *SET*s.

SHAMELLA: *Let's try it. The card that completes the first vertical SET is 1 Red Solid Oval.*

TYLER: *Okay. Wait! I get the same card for the second vertical SET!*

ERIN: *Me too! Whoa, that means our six cards are a triple interset.*

TYLER: *I thought these three new cards had to form a SET. What's up with that?*

SHAMELLA: *Well, I suppose technically, three copies of the same card could pass as a SET—an "all four attributes the same" SET.*

ERIN: *Okay, but still, what's up with that? Like, when we're left with six cards at the end, how often will we end up being able to pair them in such a way as to get a triple interset?*

Erin asks an interesting question. We will get an answer using computer simulations in chapter 10. The results suggest that this happens about 18% of the time when there are six cards left.

---

[13] This stands for *quod erat demonstrandum*, which is Latin for "which is what had to be proved." Traditionally, proofs in mathematics end with QED. Various teachers have also claimed these letters stand for "quite easily done," or more accurately, "quite enough, dammit."

Figure 4.12. Five cards.

One way to simplify the search for pairings that would give a triple-interset card is to notice that certain configurations force the cards to be paired in certain ways. For example, if the six cards include four reds, one green, and one purple, then the color of the triple-interset card must be red. This means we would need to pair the purple and green cards. Similar reasoning applies to the other attributes, of course. For instance, if there are three ovals and three diamonds among the six cards, the triple-interset card must be a squiggle, and each pair must include an oval and a diamond. You can practice these ideas with the six cards from figure 4.11.

Application 7: A five-card exception.

Suppose we are playing the End Game, and there are five cards remaining. Is there any restriction on those five cards? In other words, can *any* five cards be left at the end of the game?

Try playing the End Game with the five cards in figure 4.12.

You should have found that the missing card is 1 Green Solid Oval. But wait—*that card is already in use!*

We conclude that there are collections of five cards that *cannot possibly* be the cards left at the end. These particular collections of exceptional configurations have a very special geometric structure, which we will explore in chapter 5.

4.6 WHAT'S SO SPECIAL ABOUT THE NUMBER 3?

What if the inventor of SET had decided that each attribute would have four choices, instead of three? What would happen if we tried to play this game? Consider a deck where each attribute has four expressions, like Edna's deck in the chapter 2 exercises, with $4^4 = 256$ cards, and where *SET*s comprise four cards. Our mantra isn't changing—Number, Color, Shading, Shape—but we're adding another expression of each

TABLE 4.3.
Assignment of coordinates to cards.

Attribute	Value		Coordinate
Number	4, 1, 2, 3	↔	0, 1, 2, 3
Color	green, purple, red, brown	↔	0, 1, 2, 3
Shading	empty, striped, checkered, solid	↔	0, 1, 2, 3
Shape	diamonds, ovals, squiggles, rectangles	↔	0, 1, 2, 3

of those attributes. For number, we add the number 4; for color, let's add brown; for shading, we could add a checkered pattern; and for shape, let's add rectangles. Now we have four expressions of each attribute.

Let's see what our coordinates would look like (see table 4.3). And let's bring back our students.

SHAMELLA: *I guess now we're working mod 4.*

ERIN: *That means that in order to make a SET, we need to find four points, and each of their coordinates needs to sum to 0 (mod 4).*

TYLER: *Let's pick a SET to see if this works. To make this easy, we should pick a SET that differs in only one attribute: 1 Purple Checkered Rectangle, 2 Purple Checkered Rectangles, 3 Purple Checkered Rectangles, and 4 Purple Checkered Rectangles.*

SHAMELLA: *Sounds like a SET to me!*

ERIN: *Translating to coordinates using that new table, we get (1,1,2,3), (2,1,2,3), (3,1,2,3), and (0,1,2,3). Now we need to add the coordinates.*

TYLER: *Okay! First coordinate: 1+2+3+0=6.*

SHAMELLA: *Second coordinate: 1+1+1+1=4.*

ERIN: *Third coordinate: 2+2+2+2=8.*

TYLER: *Fourth coordinate: 3+3+3+3=12.*

SHAMELLA: *Now to make sure this is a SET, we need to translate these sums into numbers modulo 4. Starting with the first coordinate, $6 = 2$ (mod 4), $4 = 0$ (mod 4), $8 = 0$ (mod 4), and $12 = 0$ (mod 4), so the sum will be $(2, 0, 0, 0)$ (mod 4).*

ERIN: *Wait! Isn't the sum supposed to be (0,0,0,0)? What's that 2 doing there?*

TYLER: *Did we mess up?*

SHAMELLA: *No, our arithmetic is fine, and that SET you picked should have been fine too. And the latter three coordinates worked, because those were the attributes that were the same.*

ERIN: *That makes sense, because when the attributes are the same, they're assigned the same number, and as we keep seeing, any multiple of n is equivalent to 0 (mod n), so any number multiplied by 4 will always be 0 (mod 4). The coordinate that's not 0 is the one where the attribute expressions were all different.*

TYLER: *Yeah, because we had 0, 1, 2, and 3. But when we add those we get 6, which is not equivalent to 0 mod 4. What does this mean?*

SHAMELLA: *It means that some of the modular arithmetic facts we've been discussing were specific to the number 3, and will no longer apply to this version of the game. Here's another problem. If I pick two cards, then there is no longer a unique card that completes the SET. For example, if I take the cards with coordinates (0,1,2,3) and (1,2,3,0), then each attribute is different, so this will be an all-different SET.*

ERIN: *I see. I could take (2,3,0,1) and (3,0,1,2) as the last two cards, but I could also take (3,3,1,1) and (2,0,0,2) as the last two cards. There are lots of other choices that would work, too.*

TYLER: *So there really is something special about the number 3.*

ERIN: *Obviously! It's my favorite number.*

Shamella is correct. *SET*s in this game (where there are four expressions for each attribute) will not sum to (0,0,0,0), and this is a big problem. But this isn't the only problem with this game. It will also be the case that groups of four cards could sum to (0,0,0,0) without being a *SET*. For instance, in a single attribute, we could have coordinates 0, 0, 2, and 2. This violates the all-the-same-or-all-different rule of SET, but $0 + 0 + 2 + 2 = 4 = 0$ (mod 4).

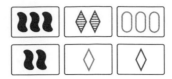

Figure 4.13. Exercise 4.1.

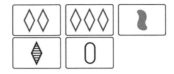

Figure 4.14. Exercise 4.3.

Executive Summary: When it comes to SET, 3 really is the magic number.

EXERCISES

EXERCISE 4.1. Your friend Chubbles has played a game of SET and claims the six cards in figure 4.13 were left at the end. Assuming no mistakes were made in the play of the game, explain why Chubbles is a rotten liar.

EXERCISE 4.2. This references exercises 2.2 and 2.3. One problem with Edna's version of SET is that it messes up the beauty of modular arithmetic that we get from working mod 3. Stefano thinks his version of the game with the added attribute of "feel" *would* work. Is Stefano right or wrong?

EXERCISE 4.3. Play the End Game! Suppose the five cards in figure 4.14 remain in playing the game.

a. Find the missing card.
b. Explain why the missing card cannot form a *SET* with any pair of cards that remain.

EXERCISE 4.4. Here's a larger End Game problem. Find the missing card in figure 4.15 and determine if it makes a *SET* with two of the cards.

EXERCISE 4.5. When you are given two cards, there is a unique third card that completes the *SET*.[14] When we use coordinates for this, if we are

---

[14] Everyone knows that by now!

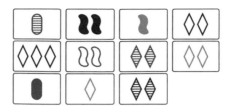

Figure 4.15. Exercise 4.4.

given cards A and B, we need to find coordinates for a card C satisfying $C = (0, 0, 0, 0) - A - B$. Is it possible for this method to mess up? That is, is it possible to have $C = A$ or $C = B$?

EXERCISE 4.6. We saw in exercise 2.4 that the center of an interset is unique. Explain why this is true using coordinates and modular arithmetic. That is, if you have four cards A, B, C, and D, and if ABX and CDX are both *SET*s for some card X, then it's impossible for ACY and BDY to be *SET*s for some other card Y.

EXERCISE 4.7. We used the midpoint formula to find the third card that makes a *SET* with given cards A and B. This gives $C = (A + B)/2 = 2A + 2B$. Show that if $A + B + C = (0, 0, 0, 0)$ (mod 3), then $A = 2B + 2C$, $B = 2A + 2C$, and $C = 2A + 2B$.

EXERCISE 4.8. In this chapter, we found that the number of expressions of each attribute at the end of the game must be the same modulo 3.

a. Verify that when there are 6 cards left at the end, the only possibilities are {4, 1, 1}, {2, 2, 2}, {3, 3, 0}, and {6, 0, 0}.
b. What are the possible numbers of expressions of attributes when there are 9 cards left at the end?
c. What are the possible numbers of expressions of attributes when there are 12 cards left at the end?

EXERCISE 4.9. Three cards form a *SET* precisely when the sum of the coordinates is $(0, 0, 0, 0)$ (mod 3). This shows that three cards summing to $(0, 0, 0, 0)$ does not depend on how we assign coordinates to the cards.

a. Show that this is false for two cards—find an example of two cards that sum to $(0, 0, 0, 0)$ using one coordinate assignment, but not for a different assignment.
b. Repeat for four cards.

EXERCISE 4.10. Suppose we play the game with four expressions for each attribute, as in table 4.3.

a. Suppose we choose the following two cards: 1 Purple Checkered Squiggle and 2 Brown Checkered Rectangles. Find two different pairs of cards that complete the *SET*.
b. Suppose the first two cards differ in all four attributes. How many different *SET*s contain these two cards?
c. Find four cards in this version of the game that sum to (0,0,0,0), but do not form a *SET*.
d. Repeat parts (b) and (c) for the game with five expressions for each attribute.

PROJECTS

PROJECT 4.1. This project uses ideas from chapters 3 and 4. Suppose we choose six cards at random from the deck. What is the probability these six cards can be the six cards left at the end of a game of SET?
To get an answer, we'll need to do a few calculations. First, we need to figure out how many ways six cards $\{A, B, C, D, E, F\}$ can be chosen so that $A + B + C + D + E + F = (0, 0, 0, 0)$. Once we have that number, we'll remove the configurations consisting of two *SET*s, since the game doesn't end if *SET*s remain.

a. There are $\binom{81}{5}$ ways to choose five cards, and once these five cards are chosen, the sixth is determined by the equation $A + B + C + D + E + F = (0, 0, 0, 0)$ (mod 3). Among these five-card selections, we need the number of *bad* configurations where the sixth card matches one of the previous five. Suppose E is a "bad" card for $\{A, B, C, D, E\}$, i.e., when we use the End Game calculation to determine the sixth card, we get E. Show that $A + B + C + D = E$. This can happen $\binom{81}{4}$ ways.
b. Show that the number of bad five-card configurations is $\binom{81}{4} - 78 \times 1080$. [Hint: The $\binom{81}{4}$ ways to choose four cards A, B, C, and D include situations when the collection $\{A, B, C, D\}$ includes a *SET*. Explain why these configurations need to be eliminated from the count.]

c. Now explain why the total number of six-card configurations that sum to (0,0,0,0) equals

$$\frac{\binom{81}{5} - \left(\binom{81}{4} - 78 \times 1080\right)}{6} = 4{,}007{,}016.$$

d. Finally, to get the number of six-card configurations that can occur at the end of the game, we need to eliminate situations where the six cards consist of two disjoint *SET*s. Show that the final answer is

$$4{,}007{,}016 - \frac{1080 \times 962}{2} = 3{,}487{,}536.$$

e. Conclude that the probability that six randomly chosen cards could be the six cards at the end of a game is approximately 1.07%.

We can estimate the answer we got in part (c) above using probability. Here are two quick and dirty ways to do this:

f. Choose six cards at random in $\binom{81}{6}$ ways. Then the sum of the first coordinates of the six cards is (approximately) equally likely to be 0, 1, or 2 (mod 3). The same is true for the second, third, and fourth coordinates. Conclude that the number of configurations that sum to (0,0,0,0) is approximately $\binom{81}{6}/81$.

g. Alternatively, note that if we choose any five cards from the deck, each of the 81 cards in the deck is equally likely to be the sum of the five cards chosen. Show that this argument gives $\binom{81}{5} \times \frac{76}{81} \times \frac{1}{6}$ ways to get six cards that sum to (0,0,0,0).

h. Show that the answers to (f) and (g) are identical, and figure out how close these answers are to the exact value found in part (c). Then take a break—maybe get a bite to eat.

SET and Geometry

5.1 INTRODUCTION

All of your friends are busy playing SET, so you've cloned new friends. Specifically, you have cloned three ancient scholars, Socrates, Euclid, and Theano, who are now in your home discussing geometry. Socrates was a classical Greek philosopher famous for his technique of posing a series of questions (rather than simply lecturing or stating facts) to help students come to their own realizations, now often referred to as the Socratic method.

Euclid is widely considered the father of geometry as well as mathematical rigor and is famous for axiomatizing geometry. His book *The Elements* is considered to be the most important mathematics book ever written. Euclidean geometry is most often taught in high school using axioms and theorems and proofs. (Mathematical proofs could actually be taught in any branch of mathematics, but it's thanks to Euclid that high-school students tend to associate proofs with geometry.)

Theano was a philosopher and mathematician who ran the Pythagorean school following the death of Pythagoras. She may also have been the wife of Pythagoras, or the daughter of Pythagoras, but very little is known about her. This is not surprising, given that she lived during a time when women were considered property and were generally not allowed to receive an education. Despite these restrictions, she was an avid scholar and a prolific writer. She is said to have written many texts on such diverse topics as physics, astronomy, psychology, and medicine, but her most important work was a text on the golden ratio.

SOCRATES: *I say, how grand it is to be alive again, and on such a fine day! My fellow scholars, I understand we are meant to be discussing the branch of mathematics called geometry. I must therefore insist on beginning by posing the crucial question: What is geometry?*

EUCLID: *Well Socrates, if you speak ancient Greek—which you do—then you must know that "geometry" literally means "earth measurement." It is the study of shapes and sizes, and of the general properties of space.*

THEANO: *Yes, quite. Traditionally, "geometry" has referred to the study of physical space, the "measurement" of the "earth." However, there are branches of geometry that are more abstract, describing spaces that can be completely imaginary. Some of these are called non-Euclidean geometries.*

SOCRATES: *My word, is that so? Do you mean to say that there are types of geometry of which Euclid never conceived?*

EUCLID: *Evidently. All of the work I did on geometry was later titled Euclidean geometry to distinguish it from more abstract geometries. Euclidean geometry was my attempt to describe the physical spaces we encounter in our daily lives, and it is all built upon five axioms.*

SOCRATES: *Do tell. And what precisely is an axiom?*

THEANO: *An axiom—also known as a postulate[1]—is a statement that we take to be a given, an unprovable yet indisputable truth. We can think of axioms as the building blocks that we use to prove theorems. We have to start somewhere—we can't prove anything without a set of statements we accept to be true.[2] And thus, each branch of geometry has its own set of axioms, or put another way, each set of axioms gives rise to a unique branch of geometry.*

---

[1] Our scholars will use the words "axiom" and "postulate" interchangeably.

[2] Consider the creation of a dictionary: Each word is defined in terms of other words. This produces undefined cycles of words, a fact we conveniently ignore most of the time by accepting some of the most basic words of our language as essentially axiomatic. For example, most people would probably not argue about the definition of the word "it."

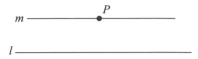

Figure 5.1. The parallel postulate. There is a unique line *m* through the point *P* parallel to the given line *l*.

SOCRATES: *Most intriguing. Might you share with us some examples of your axioms, Euclid?*

EUCLID: *I am glad you asked. My first axiom states that a line segment can be drawn connecting any two points, and my second axiom states that any line segment can be extended infinitely in both directions to create a line. Taken together, these postulates indicate that two points determine a unique line.*

SOCRATES: *And indeed they must, for I can imagine no other possibility. But this raises a question: What do axioms look like in non-Euclidean geometry?*

THEANO: *The main axiom that changes from geometry to geometry is the fifth Euclidean axiom, also known as the parallel postulate. One way to state it is to say that if I have a line and a point not on the line, then there is exactly one line through that point that is parallel to my original line. In figure 5.1, I've drawn a picture for your edification.*

There are perfectly valid geometries where this axiom no longer holds.[3]

EUCLID: *The statement Theano just gave is most likely the statement of the parallel postulate that our readers would have learned in school. However, my original statement of this postulate, while logically equivalent, was a bit more complicated: "If a straight line crossing two straight lines makes the interior angles on the same side less than two right angles, the two straight lines, if extended indefinitely, meet on that side on which the angles are less than the two right angles." I've included my own illustration in figure 5.2.*

---

[3] In elliptic geometry, there are no parallel lines, and in hyperbolic geometry, there are infinitely many lines parallel to a given line.

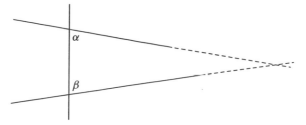

Figure 5.2. Euclid's version of the parallel postulate. If $\alpha + \beta$ is less than two right angles, then the two lines meet on the same side.

SOCRATES: *Now this is interesting, because that statement you just gave sounds much more complicated than your initial axioms. In fact, it sounds to me like it needs to be proved, yet Theano said earlier that axioms are unprovable. How is it possible that this statement needs no proof?*

EUCLID: *Excellent question, Socrates. There does seem to be something different about this axiom, and in fact, I delayed introducing it in my books as long as possible.[4] And even though I introduced the parallel postulate as an axiom, people actually tried for a very long time to prove it as a theorem using the other axioms. However, this is impossible, since non-Euclidean geometries exist, meaning a consistent geometry can arise that does not satisfy that postulate.*

THEANO: *Yet despite the confusion surrounding this axiom, it turns out that some of the attempts to prove the parallel postulate did produce important mathematics.*

Theano is correct (somehow, even though this happened long after she lived): two nineteenth-century mathematicians, Bolyai and Lobachevsky, independently discovered that you could have a consistent geometry satisfying the first four axioms but not the fifth. (Gauss claimed he had done the same, but he never published this work.) Elliptic and hyperbolic geometry are examples of consistent geometries that satisfy Euclid's first four postulates but not the fifth, the parallel postulate. They have their own versions of a parallel postulate. So,

---

[4] Of course, we don't know whether Euclid tried to prove it and couldn't, or whether he knew it was an axiom. We embrace the mystery.

"Euclidean" geometry is what we now call the geometry in which the parallel postulate holds.

5.2 FINITE AFFINE GEOMETRY

THEANO: *Let us now discuss a specific, non-Euclidean branch of geometry called finite affine geometry.*

SOCRATES: *With pleasure, Theano, and I must once again insist on beginning by posing the crucial question: What is finite affine geometry?*

EUCLID: *I never considered the possibility of finite geometries, because I intended geometry to be a reflection of the world around us, which I perceive as infinite. But I suppose a finite geometry is one that contains a finite number of points.*

THEANO: *Undoubtedly. In finite geometries, the third and fourth Euclidean axioms—which essentially define circles and right angles—do not apply. Instead, these geometries are all about the incidences holding between finite sets of points and lines. Finite affine geometry deals exclusively with points, lines, planes, and hyperplanes.[5]*

SOCRATES: *Am I to understand that there are branches of geometry in which circles and right angles do not exist? What could be the purpose of a geometry that deals only with points and lines?*

THEANO: *The purpose, as I understand it, is that points and lines in this context are more abstract than literal points and lines. Instead, points and lines can represent anything. For example, maybe points are humans, and lines are relationships between humans.*

EUCLID: *What a novel concept!*

SOCRATES: *Forsooth, this revelation unleashes a world of possibilities. I wonder, Theano, to what other settings can we generalize finite affine geometry?*

---

[5] The term "hyperplane" can mean a variety of things depending on context. We'll see it defined with respect to finite affine geometry later on in this chapter.

THEANO: *Many, Socrates, but there is one setting of particular interest to us here. Finite affine geometry turns out to be an excellent model of the card game* SET.

EUCLID: *How fortuitous that* SET *is the subject of this very book.*

THEANO: *Indubitably. The axioms for finite affine geometry translate remarkably well into the rules of the card game.*

SOCRATES: *And thus, I must now ask, what are the axioms for finite affine geometry?*

To answer this question, we will begin by working in two dimensions. Over the course of the rest of this chapter, we will work our way up first to three dimensions and finally to the four dimensions that comprise the SET deck.

Finite Affine Plane Axioms

Axiom 1. *There are at least three non-collinear[6] points.*

Axiom 2. *Every line contains at least two points.*

Axiom 3. *Two points determine a unique line.*

Axiom 4. *For any line l and any point P not on l, there is exactly one line containing P and not containing any point on l. This line is said to be parallel to l.*

We can make some comments about these axioms:

- The purpose of axioms 1 and 2 is to avoid "boring" geometries, for example, situations where all the points lie on a single line, or where lines consist of a single point.
- Axiom 3 is essential to any geometric structure.
- Axiom 4 is one way of stating the parallel postulate, and is illustrated in figure 5.1. There are others, but this one works best for our purposes. We need to be working in a plane for this to make sense, though. In three-dimensional space, there are *skew* lines, i.e., lines that don't intersect, but that aren't parallel.

---

[6] Non-collinear simply means "not on the same line."

What Does This Have to Do with SET?

The axioms for finite affine planes apply beautifully to SET. Think of the cards as points in our geometry, and think of *SET*s as lines. With this substitution, let's take another look at the axioms.

Axiom 1. There are at least three cards that are not in the same SET.
SET interpretation: Self-evident.
Axiom 2. Every SET contains at least two cards.
SET interpretation: Self-evident.
Axiom 3. Two cards determine a unique SET.
SET interpretation: Every pair of cards determine a unique *SET*. We've been calling this the "fundamental theorem of SET."
Axiom 4. For any SET and any card not in the SET, there is exactly one SET containing this card that is parallel to the original SET.
SET interpretation: Given a *SET* and a card not belonging to that *SET*, there is a unique *SET parallel* to the given *SET*. To make sense of this connection to the game, we will need to define "parallel" in the SET universe. This will happen soon, we promise.

EUCLID: *To think that seemingly abstract axioms can translate into rules for a popular card game! I only wish we had had SET in my day.*

THEANO: *We all do, Euclid. Not to mention computers. In any case, now that we have axioms, we have a reasonable geometry.*

SOCRATES: *Well yes, apparently we do, but I must admit that until now I have been relying on faith. This geometry seems conceivable, but what does it actually look like?*

Using just the four axioms, here's what we can do:

1. We can prove that all lines have the same number of points. This is reasonably entertaining to go through, and it is also the point of exercise 5.1

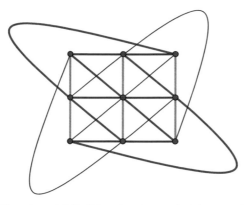

Figure 5.3. The affine plane AG(2, 3) has 9 points and 12 lines, with 3 points on each line.

2. If we then assume each line has exactly three points, we can prove that there is only one point–line structure that satisfies all the axioms. This structure will consist of 9 points and 12 lines.[7] This is done in the rather long, detailed exercise 5.2. See figure 5.3. (It's pretty!)

Mathematicians call this plane AG(2, 3). The letters AG tell us this is an affine geometry, and (2, 3) tells us we're working in 2 dimensions, with 3 points on each line.

5.3 THE PARALLEL POSTULATE AND SET

Now for the fun part. Our geometers have obtained a deck of SET cards.

SOCRATES: *How do we draw a plane using the* SET *cards? And what does "parallel" mean in the context of* SET?

THEANO: *Those are two separate questions, Socrates, but we have actually already seen the answer to the first question in*

---

[7] Our affine plane has three points per line. Other numbers are possible, but there are restrictions. It is known that if $n = p^k$, where p is a prime number and k is a positive integer, then there are affine planes with n points per line. No other numbers are known to work, but finding precisely which numbers are possible is a famous open problem.

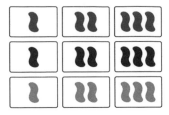

Figure 5.4. The affine plane AG(2, 3). SETs occupy the same positions in this picture that lines do in figure 5.3.

chapters 1 and 2. Let us take all the solid squiggles and arrange them as in figure 5.4, making a plane of SET cards similar to the ones we have previously seen. Then it is clear the lines in the geometry in figure 5.3 match the location of the SETs in figure 5.4.

EUCLID: *This is most astonishing! In any geometry, two lines are "parallel" if they are in the same plane and they don't intersect. Looking at figure 5.3, we see four collections of lines: horizontal, vertical, and two types of diagonal[8] lines, and each collection of parallel lines is colored the same.*

SOCRATES: *Incredible. I believe I am beginning to understand what could make SETs "parallel" to one another. For instance, is it true that each horizontal SET is parallel to the other horizontal SETs in figure 5.4, just as the horizontal lines in figure 5.3 are parallel?*

THEANO: *Precisely, and the same is true for each collection. For example, the two SETs in figure 5.5 are parallel. These correspond to the SETs in figure 5.4 that correspond to two of the blue diagonal lines in figure 5.3.*

---

[8] You've probably noticed that four of the diagonal lines are curved. Remember: This is an abstract geometry! We can bend lines if we need to, in order to demonstrate their existence connecting points. You may also notice that the "curves" seem to cross each other, but these are not actual points of intersection—the only valid points in this geometry are ones we already defined, shown in red. This may violate your intuition at first, since you may be used to thinking in Euclidean geometry, so you may want to give yourself some time to get used to these new axioms.

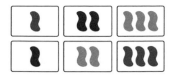

Figure 5.5. These two *SET*s are parallel.

SOCRATES: *Could one perhaps say that the 12 SETs in figure 5.4 break into four classes, with three parallel SETs in each class?*

THEANO: *One certainly could. That is an elegant summary.*

This will work for any plane we can construct. For instance, back in chapter 2, we showed how to complete a plane using three cards that don't form a *SET*. Figure 5.3, the abstract affine plane, tells us which *SET*s are parallel to each other. Since these configurations consist of 9 cards and 12 *SET*s, they are quite special. They are called *magic squares* on the Set Enterprises website, because for every pair of cards in the square, we can find the card that completes their *SET* within the square (this is the "closed" property we discussed in section 1.2). We will continue to refer to such structures as planes, since we are discussing geometry. This may be strange to you at first, because Euclidean planes are infinite, so take a moment to get used to the idea that a "plane" in our universe is a finite set of 9 points (cards) containing 12 lines (*SET*s).

Something else to notice about figure 5.4 is that we isolated two attributes: the cards are all solid squiggles. In a sense, attributes correspond to dimension. Right now we are dealing with a two-dimensional structure, i.e., a plane, and the easiest way to do this is to "fix" two attributes (shape and shading), allowing only the other two attributes (number and color) to vary. The number of attributes that can vary corresponds to dimension.

But this is not the only way to construct a plane—we can use all four attributes, and the plane will still be "closed" (the card completing the *SET* of any pair of cards in the plane will be contained in the plane), as in figure 5.6. This works because our definition of parallel doesn't change as we add dimensions (attributes), which will be important soon.

EUCLID: *Fellow geometers, I would like to revisit the parallel postulate for SET.*

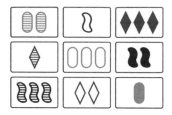

Figure 5.6. Another plane, with all four attributes represented.

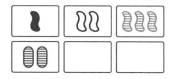

Figure 5.7. A *SET* and a card. Find the *SET* that contains the card and is parallel to the original *SET*.

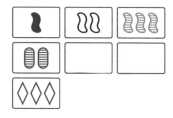

Figure 5.8. Adding a card to make a vertical *SET*.

THEANO: *Now would be a good time to do this, since we have just seen some examples of parallel SETs.*

SOCRATES: *Naturally, for there remains an unaddressed question: If I am given a SET and a card not in the SET, how can I find the unique SET parallel to the first SET?*

To satisfy our scholars, let's pick a *SET* and a card not in the *SET*, as in figure 5.7.

First, complete the vertical *SET* in the first column, as in figure 5.8.

As in chapter 2, we continue to add cards to this plane—see figures 5.9 and 5.10. We will not need to finish the entire plane. (But don't let that stop you from finishing it if you'd like to!)

For the last step, complete the *SET* in the second row, and you can now get rid of the extra card in the left vertical *SET*, as in figure 5.10. This *SET* will be parallel to the original *SET*.

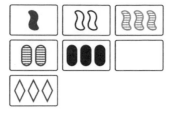

Figure 5.9. Adding a card to complete a diagonal *SET*.

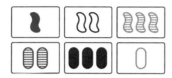

Figure 5.10. Adding the card to complete the *SET* parallel to the given *SET*.

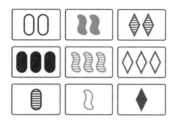

Figure 5.11. A plane with four collections of parallel *SET*s: three horizontal, three vertical, three diagonal with "positive slope" (lower left to upper right), and three diagonal with "negative slope" (upper left to lower right).

SOCRATES: *That is quite illuminating, but how would we be able to determine if two SETs are parallel without determining that additional card, 3 Red Empty Diamonds?*

Socrates asks a great question. (That's what he does.) Can we tell whether or not two *SET*s are parallel by just looking at the *SET*s? The answer is yes. Take a look at some of the parallel *SET*s, and see if you can find some pattern based on the examples. We've given you another plane in figure 5.11 to help you figure it out. (Remember, horizontal *SET*s are parallel, vertical *SET*s are parallel, etc.)

Our solution to Socrates' last question is all about pattern recognition. Before giving you the solution, we will first need a brief digression on *cycles*.

Cycles: A cycle of numbers (or colors, or objects of any kind) is a sequence placed in circular order. For instance, if you seat three people around a circular table (and label them 1, 2, and 3), then going around the table clockwise (several times), you might encounter the people in the order $1 \rightarrow 2 \rightarrow 3 \rightarrow 1 \rightarrow 2 \rightarrow 3 \rightarrow 1 \cdots$. We will write this cycle as (1, 2, 3). The last number is assumed to cycle back to the first one. This means that the cycles (1,2,3), (2,3,1), and (3,1,2) are equivalent—they correspond to the same seating arrangement, moving clockwise around the table.

How many ways can three people sit around a table? While there are $3! = 3 \times 2 \times 1 = 6$ different ways in which we could order the numbers 1, 2, and 3, there are only two distinct 3-cycles.[9] The other three ways of ordering the numbers—(3,2,1), (1,3,2), and (2,1,3)—are also all equivalent to each other, and in fact, they represent the (1,2,3) cycle going in reverse (or moving counterclockwise around the table).

Staring at *SET*s that are parallel and *SET*s that aren't (see figure 5.12 for two *SET*s that are not parallel), we (eventually) see that for *SET*s to be parallel, we have the following conditions:

- If an attribute is the same in one *SET*, then it's the same in the other. In figure 5.10, shape is the same: the three cards in the *SET* in the top row are all squiggles, and the three cards in the parallel *SET* in the second row are all ovals.
- If an attribute is all different in one *SET*, then it's all different in the other. More importantly, in this case, you can place the cards in the two *SET*s in some order so that, for all the attributes that are not the same, the *cyclic ordering is the same*. Here's what we mean (refer back to figure 5.10):

  - Number: Moving left to right, the cards in the first *SET* cycle (1, 2, 3) and the cards in the second *SET* cycle (2, 3, 1), which is equivalent to (1, 2, 3). These are the same left-to-right cyclic order.

---

[9] In general, there are $(n - 1)!$ ways to seat n people around a circular table.

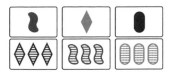

Figure 5.12. Two *SET*s that are not parallel.

- Color: Here, the two *SET*s again cycle the same way: (red, purple, green).
- Shading: Again, the two *SET*s cycle the same way: (solid, empty, striped) and (striped, solid, empty), which are equivalent.

Let's look at this cyclic condition for two *SET*s that are not parallel, the *SET*s in figure 5.12. Number is the same for each *SET* (the top *SET* is all 1s, and the bottom *SET* is all 3s) and so is shading (all solid in the top *SET*, and all striped in the bottom *SET*). That means we need to check color and shape for cyclic consistency.

- Color: Both *SET*s share the (red, green, purple) cycle.
- Shape: Moving left to right, the first *SET* cycles (squiggles, diamonds, ovals), but the second cycles (diamonds, squiggles, ovals). These cycles are different.

We conclude the two *SET*s in figure 5.12 are not parallel.

Are we sure we're right? Yes, but it's always a good idea to check your work. If we start completing *SET*s by taking a card in the top row and another in the second row, we will quickly need more than nine cards. In particular, this procedure forces us to add 2 Green Empty Ovals, 2 Green Empty Squiggles, 2 Green Empty Diamonds, 2 Red Empty Squiggles, 2 Red Empty Diamonds,...; in fact, we get every card with two empty symbols. So those two *SET*s can't be in a plane, so they can't be parallel. (In fact, we've constructed a *hyperplane* containing the original two *SET*s. We'll have more to say about hyperplanes later in this chapter.)

Finally, we can use this idea to answer Socrates' earlier question a different way: find the *SET* parallel to the *SET* in figure 5.13 (this was the top *SET* in figure 5.12) containing the card 3 Purple Striped

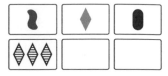

Figure 5.13. A *SET* and a card. Find the *SET* that contains the card and is parallel to the original *SET* using the cyclic procedure.

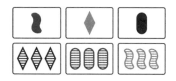

Figure 5.14. A *SET* and the parallel *SET* containing the card.

Diamonds (the left card of the bottom *SET* of same figure). Here's how we can create a parallel *SET* using the cyclic procedure instead of completing a plane.

- Number: All the cards in the *SET* are 1s, so the cards in the parallel *SET* will all be 3s.
- Color: Color cycles (red, green, purple) in the *SET*, and, since the card is purple, the equivalent cycle is (purple, red, green).
- Shading: The first *SET* is all solid, so the cards in the parallel *SET* will all be striped.
- Shape: The shapes in the *SET* cycle (squiggles, diamonds, ovals), so our parallel *SET* will cycle (diamonds, ovals, squiggles).

This means that the next card in the *SET* is 3 Red Striped Ovals, and the third card is 3 Green Striped Squiggles. That *SET* is pictured in figure 5.14.

SOCRATES: *This cyclic procedure is most interesting, but it raises a question: Does the cyclic procedure depend on the order of the cards?*

Another good question, Socrates. While it looks like the definition did depend on order, all we need for the *SET*s to be parallel is that there is *some* order where all of this works. In fact, two *SET*s are parallel if *either* all attributes that are different cycle the same way *or* they all cycle the opposite way. See the nonparallel *SET*s in figure 5.12: color

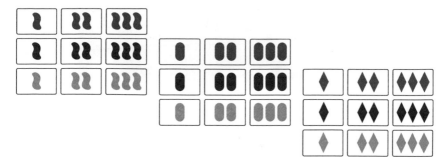

Figure 5.15. AG(3, 3) with the cards.

cycled the same way in the two *SET*s, but shape cycled the opposite way. If color had *also* cycled the opposite way, as shape does, then the *SET*s would be parallel, and we would be able to reorder the cards to make both of those attributes cycle the same way in both *SET*s.

Takeaway Message: Parallelism is a property of the *SET*s, not of the ordering.

Why is the cyclic description of parallel *SET*s equivalent to the one we obtained from the parallel postulate? Our explanation uses vectors, and it appears in chapter 8. Finally, we note that our descriptions of parallel *SET*s do not depend on dimension. This means our notion of parallel *SET*s will still work when we increase the dimension. And since this is what we're about to do, this is good news.

5.4 THREE-DIMENSIONAL AFFINE GEOMETRY: AG(3, 3)

Having mastered the affine plane, we're ready to move from two dimensions to three. The three-dimensional affine geometry whose lines all contain three points is AG(3, 3); we will call a collection of cards that form AG(3, 3) a "hyperplane," as this is the term mathematicians use. (If you liked the term "magic square," you could potentially think of a hyperplane as a "magic cube," but we won't be using that term.)

What does a hyperplane look like? It consists of three parallel planes. Imagine stacking three planes on top of each other in three layers to make a three-dimensional cube. Figure 5.15 shows a hyperplane,

Figure 5.16. A model of a hyperplane as a cube.

TABLE 5.1.
Hyperplane counts.

# cards	# SETs	# planes
27	117	39

and figure 5.16 is a photo of a three-dimensional SET "cube" in which SET cards are represented by balls of clay with the appropriate symbols.[10]

How many *SET*s are there in a hyperplane? How many planes are there in a hyperplane? Although the hyperplane is constructed from three parallel planes, it contains more than those three planes. To satisfy your curiosity, we give the answers to these counting questions in table 5.1. All of these counts will be justified more thoroughly in chapter 6.

It's a good exercise in visualization to try to locate as many of the 39 planes in the hyperplane of figure 5.15 as you can. It's easy to spot the

---

[10] Some of us are good with clay. Thanks, Liz!

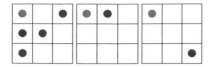

Figure 5.17. AG(3, 3) with three lines highlighted.

three planes determined by shape: the squiggles form a plane, as do the ovals, and the diamonds. But we could also take the top row of each of these planes to make another plane (i.e., all of the red cards).

More generally, to create a plane, we could choose any *SET* from the squiggles, then choose a parallel *SET* from the ovals. This would then determine the remaining cards, which would be another parallel *SET* from the diamonds. We have 12 choices for the *SET* we choose from the squiggles, then three choices for the parallel *SET* from the ovals, so we get 36 planes in this way. Adding in the 3 planes above (all the squiggles, all the ovals, and all the diamonds), it turns out that's all of them, for a total of 39.

Where are the 117 *SET*s in AG(3, 3)? We give a schematic grid-like picture in figure 5.17 to help clue you in. There are three types of *SET*s:

- *SET*s entirely contained in one of the three parallel planes. For instance, in figure 5.15, the *SET* that includes 1 Green Squiggle, 2 Purple Squiggles, and 3 Red Squiggles lives in the first plane. This *SET* is represented in figure 5.17 as three red dots. There are 12 *SET*s in each plane and 3 planes, so we have a total of $12 \times 3 = 36$ *SET*s of this first type.
- *SET*s that use one card from each of the three planes. There are two ways this can happen, and we again use the cards in figure 5.15 to illustrate.
  - We could choose 1 Red Squiggle, 1 Red Oval, and 1 Red Diamond. This is represented by the green dots in figure 5.17. Note that these cards occupy the same relative position in each of the three planes.
  - We could also choose 1 Purple Squiggle, 2 Red Ovals, and 3 Green Diamonds. This is represented by the blue dots in

figure 5.17. This time, if we overlay the relative positions the cards occupy in a single 3×3 grid, we get one of the diagonal *SET*s in a plane.

For *SET*s of this second type, if we pick a card in one plane and a card in another plane, these cards uniquely determine a *SET* (whose last card is in the third plane). Because there are 9 cards in each plane, there are thus $9 \times 9 = 81$ *SET*s of this second type, and $36 + 81 = 117$, so this accounts for all *SET*s in the hyperplane.

Notice that these possibilities for what *SET*s can look like in AG(3, 3) are very reminiscent of what defines a *SET*: all the same or all different. Either all the cards of the *SET* are in the same 3×3 subgrid, or each card is in a different one. You can say the same about the rows and about the columns.

So how do we make a hyperplane? One way is to isolate one attribute. (In fact, instructions for playing SET suggest starting with only the red cards in the deck when teaching the game to young children. We have found that this is not necessary, though.[11]) For example, all the cards in figure 5.15 are solid. We could just as easily have taken all the 1-symbol cards, or all the green cards, or all the diamonds, etc. Recall that earlier, in two dimensions, we isolated two attributes: solid squiggles. The attributes that could still vary were number and color. In this case, in three dimensions, we no longer isolate shape, so we now have three attributes that can vary: number, color, and shape. Again, the number of attributes allowed to vary corresponds to dimension, in this case three.

Just as our planes were closed (meaning the card that would complete the *SET* for every pair of cards in the plane was itself contained in the plane), our hyperplanes will also be closed. This is a property of the geometry: the only groups of cards that can be closed in this manner must be powers of 3 ($3^1 = 3$, $3^2 = 9$, $3^3 = 27$, $3^4 = 81$), since we're working in a geometry where lines are restricted to containing exactly 3 points. A *SET* (3 cards) is closed; a plane (9 cards) is closed; a

---

[11] Children pick up the full game very quickly. Then they destroy their parents.

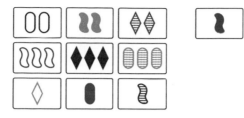

Figure 5.18. A plane and one extra card.

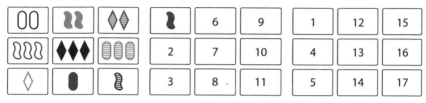

Figure 5.19. Starting with a plane and one more card, the hyperplane is completely determined. The numbers indicate one order we can use to find the unknown cards.

hyperplane (27 cards) is closed; and as we already know, the entire deck (81 cards) is closed. These are the only structures in our deck that can be closed.

Building a Hyperplane: Completing SETs

We can also make a hyperplane using a mixed collection of cards. How do we do this? Recall that a SET plane is completely determined by three cards that don't form a SET: adding one more card not in the plane will uniquely determine a whole hyperplane. So we start with a plane and a card not in the plane (see figure 5.18).

We can now build the hyperplane, card by card, where each new card completes some SET with two cards already in our collection. This depends on knowing the location of the SETs—for that, we'll rely on our grid (figure 5.17). Filling in the missing cards will have the flavor of a puzzle. There are many ways to do this—we've outlined one ordering in figure 5.19. We'll illustrate this technique by showing how to get the first three unknown cards.

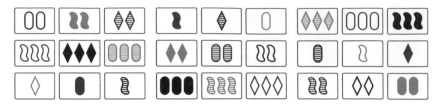

Figure 5.20. The hyperplane!

Let's fill in some of the 17 missing cards in figure 5.19.

1. The card labeled 1 in figure 5.19 completes a *SET* with 2 Purple Empty Ovals and 1 Red Solid Squiggle. So this card must be 3 Green Striped Diamonds.

2. For the card labeled 2 in the figure, we use 1 Green Empty Diamond and 3 Green Striped Diamonds (the card we placed in the first step). So card 2 becomes 2 Green Solid Diamonds.

3. For practice, we'll do one more. The card labeled 3 completes a *SET* with 1 Red Solid Squiggle and 2 Green Solid Diamonds (the card placed in step 2). This gives 3 Purple Solid Ovals. But note that this card also completes a *SET* with 3 Red Empty Squiggles and 3 Green Striped Diamonds (the first card we placed, back in step 1). We have two options for how to find this card, but they give us the same result, so we're happy.

We encourage you to fill in the remaining cards on your own. You can cheat by looking at figure 5.20, where we give the completed hyperplane.

Two things to notice:

1. There are lots[12] of different orderings of the remaining cards that we could use to complete the hyperplane. The *SET* containing the card labeled 1 must include the "extra" card—1 Red Solid Squiggle, in this case. But that leaves nine potential spots for the card labeled 1: each of the cards in the rightmost plane. Once we select a position for the label 1, we can determine any other card in the hyperplane immediately. Then we can figure out how

---

[12] This is an understatement.

many different orderings of the labels 1–17 will work in this process. See exercise 5.3.

2. When we figured out card 3 above, we had two choices for the *SET* we could have used. As we continue the process, we will have more and more *SET*s to choose from when we determine an unknown card.

For a thought experiment,[13] try this:

- How many *SET*s could we use to find the last missing card?

Given any pair of cards in a hyperplane, the card that completes the *SET* is also present. (As we have already seen, this is where mathematicians use the term "closure.") For the last missing card, we can pair up the other 26 cards (since we have a total of 27), which would give us 13 pairs. We conclude that we have 13 different ways to figure out card 17. This also shows us that each card is in 13 *SET*s.

EUCLID: *That was quite enlightening. Finite geometry mixes pattern recognition and counting with standard geometry axioms.*

SOCRATES: *And we have just witnessed examples of how this works in the affine plane as well as in three-dimensional affine space. Now, how about four-dimensional affine space, which I am to understand is where the entire SET deck lives? How can the analysis of four-dimensional affine geometry shed light on SET?*

THEANO: *Presumably, we are about to find out.*

5.5 THE ENTIRE DECK: FOUR-DIMENSIONAL AFFINE GEOMETRY AND AG(4, 3)

We are now finally ready for the geometry of the entire deck of cards. In constructing the hyperplane, we took a plane, then added one new card. Adding just one more card will define the entire four-dimensional space, i.e., the entire deck.

---

[13] Basically, all of mathematics is a thought experiment, isn't it?

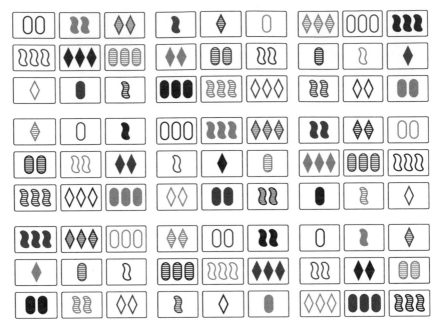

Figure 5.21. AG(4, 3) in all its glory.

From the affine geometry viewpoint, the deck of cards forms AG(4, 3), the four-dimensional affine geometry with three points per line. In constructing AG(4, 3) from the cards in the deck, we end up with an organized, somewhat mind-blowing 9 × 9 array. Each *SET*, plane, and hyperplane will be visible here.[14] Figure 5.21 is one example of such an array. We could consider this a "magic deck," or a four-dimensional hyperplane.

How could you do this yourself? You have all the tools you need. First, make a hyperplane, as you did in the previous section. Choose a card you didn't use in the hyperplane. Then, take one of the planes in your hyperplane and the extra card to make a new hyperplane. Continue, completing *SET*s like crazy to determine the missing cards.

In fact, you don't need to start with a hyperplane. To make your own beautiful array, suitable for framing, you need to place just five cards! Here's why:

[14] If you know where to look.

- First, working by analogy in lower dimensions, we know that any two cards determine a unique *SET*. We've already translated this fact about the game into geometric terms:

  Every two distinct points determine a unique line.

- When we made our first plane in chapter 1, we needed three cards that didn't form a *SET*. Then the rest of the plane was completely determined. From the geometry point of view, this translates to the following well-known fact:

  Three non-collinear points determine a unique plane.

- For the hyperplanes, we needed four cards, provided no three of them were in the same plane. Geometrically, this is equivalent to the less familiar fact from four dimensions:

  Four non-coplanar[15] points determine a unique three-dimensional hyperplane.

So we need five cards to build AG(4, 3), no four of which are in a three-dimensional hyperplane. Try making your own some day when you've got a full deck of cards, a clean floor, and some time on your hands.

We conclude our treatment of AG(4, 3) with a few remarks about the remarkable figure 5.21.

1. It's easy to see all the *SET*s parallel to a single *SET*. For example, take the *SET* consisting of 2 Purple Empty Ovals, 2 Green Solid Squiggles, and 2 Red Striped Diamonds, i.e., the top row of the plane in the top left of our display. Then there are 26 other *SET*s parallel to this *SET*, and they are easy to spot. They are just the 26 horizontal *SET*s in the nine 3 × 3 grids. We explore this in more detail in chapter 8.

2. Now, imagine shrinking each of the 3 × 3 grids to a point; if you take three of them that are in the same position as a line in

---

[15] As you might have guessed, this means "not contained in the same plane."

TABLE 5.2.
SET deck counts.

# cards	# SETs	# planes	# hyperplanes
81	1080	1170	120

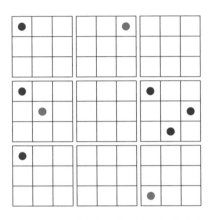

Figure 5.22. *SET*s in the "magic deck."

AG(2, 3), then those three planes will form a hyperplane! (One way to "visualize" this is to hold the picture really far from your face.)

3. Finally, there are many things to be counted here. We give the total number of *SET*s, planes, and hyperplanes in the configuration in table 5.2. All of these counts are explained in chapter 6.

Where are the *SET*s in figure 5.21? As we did for hyperplanes, we use a grid in figure 5.22 to show the relative location of the *SET*s. There are again three possibilities:

1. The *SET* is contained entirely within one 3 × 3 subgrid, where the cards are in the same position as a *SET* in a plane (blue).
2. The *SET* has one card in each of three 3 × 3 subgrids, where those subgrids are themselves in the same position as a line in AG(2, 3), and the cards are in the same position in each plane (red).

3. The *SET* has one card in each of three planes, where those planes are themselves in the same position as a line in AG(2, 3), and when you superimpose them, they are in the same position as a *SET* in a plane (green).

Executive Summary: Figure 5.21 is awesome, and it is well worth your time to stare at it and look for patterns.

5.6 MAXIMAL CAPS—A PREVIEW

While you were staring at figure 5.21, our three scholars played numerous rounds of SET, and now Socrates has a question.

SOCRATES: *I notice it is not uncommon to come across a configuration of 12 cards that do not contain a SET, but when we deal out three more, are we guaranteed to be able to find a SET in the new configuration of 15 cards? I suppose what I am really asking is, what is the largest number of cards we could have with no SETs?*

EUCLID: *As a geometer, I would ask for the largest number of points in the geometry* AG(4, 3) *that did not contain a line.*

THEANO: *And as a person who knows the answer, I would say "20."*

This is an important question, and geometers answered it before[16] the invention of SET. We return to this question in chapter 9, where we explore the structure in greater detail.

This can also have an impact on the play of the game. The rules of the game say to lay out 12 cards, and in the event that there isn't a *SET* in those 12, lay out 3 more, for a total of 15. However, it is possible (though very rare) that there is still no *SET* in those 15 cards.[17] In fact, even adding 3 more cards (bringing the total to a rather unwieldy 18)

---

[16] This fact was first proved by Giuseppe Pellegrino in 1971 in a paper titled "Sul massimo ordine delle calotte in $S_{4,3}$," long before SET was invented. It's in Italian. We have not read it.

[17] It has happened to us.

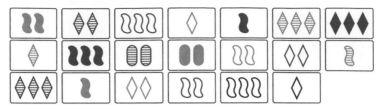

Figure 5.23. The largest cap has 20 cards. Search all you like, but you won't find any *SET*s here.

still does not guarantee the presence of a *SET*. Only after adding 3 more cards, bringing the total to (a ghastly) 21, are we certain that there is a *SET* in the layout.

Geometers call collections of points that contain no lines *caps*. What does a 20-card cap look like? We give one in figure 5.23.

For now, here's a quick preview of chapter 9 that we can't resist. We can understand the geometric structure of the cards in figure 5.23: the 20 cards can be broken up into 10 pairs, with the same card completing each of these 10 *SET*s. To see this, look at the first two cards, and complete the *SET*. Do the same thing for the next two cards, and the next pair, and so on. Every one of these *SET*s is completed by 2 Purple Empty Ovals, so the cap is a massive interset. In fact, this is a *decuple*[18] interset.

5.7 THE SIX-CARD THEOREM

Now that we've discussed maximal caps, we are ready to revisit the End Game. The cards left over at the end of the game must form a cap, since no *SET*s will be contained in those cards.

How many cards can we have left at the end? The number left must be a multiple of three, but we can't have three cards left at the end (we discussed this in chapter 1, and used modular arithmetic to explain why in chapter 4). It's possible (but rare) that there are no cards left, meaning you've cleared the whole deck.

---

[18] This is, apparently, the word for multiplying by ten, as in "double," "triple," and "quadruple," but for ten. We had to look it up.

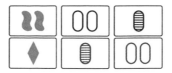

Figure 5.24. Six cards left at the end of the game played by Socrates, Euclid, and Theano.

So if there are cards left at the end, there must be at least 6. We could also have 9 remaining on the board with no *SET*s, and (infrequently) 12; it's extremely rare that more than 12 cards remain.[19]

We examined the six-card case in chapter 4, where we showed that there is some extra structure. As an application of modular arithmetic, we showed the following:

- Break up the six cards into three pairs any way you like. Add a card to each pair to complete a *SET*. Then the three cards you added form a *SET*, or the three cards you added are all the same.

This tells us there are two cases to examine when six cards remain at the end of the game:

1. The cards form a triple interset.
2. They don't form a triple interset, so every pairing produces a distinct *SET* (with no repeated cards).

We can use what we know about hyperplanes to distinguish these two cases.

Case 1. Six cards left form a triple interset.

What does the structure of the six leftover cards look like when those cards form a triple interset? It turns out these cards are contained in a hyperplane. In fact, we can use these cards to determine a unique hyperplane. Figure 5.24 shows a triple interset left at the end of a spirited game among Socrates, Euclid, and Theano. These cards all live in the hyperplane shown in figure 5.25.

Looking back at the triple interset of figure 5.24, we see that 3 Green Solid Ovals completes each *SET*. We've placed this card in the upper left

---

[19] A simulation in chapter 10 gives approximate probabilities for this.

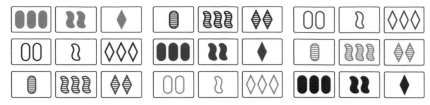

Figure 5.25. This hyperplane contains the triple interset in figure 5.24 left at the end of a game played by Socrates, Euclid, and Theano.

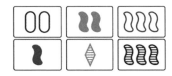

Figure 5.26. Six cards left at the end of another game played by Socrates, Euclid, and Theano. They play really fast.

corner of the hyperplane in figure 5.25. Now imagine our hyperplane is a genuine, three-dimensional cube (as in figure 5.16). We've placed our special six cards in such a way that these three *SET*s form three edges of the cube that intersect in one point (card), the card that completes the triple interset. This may remind you of the x-, y-, and z-axes in three-dimensional Cartesian space.

Takeaway Message: If the six cards form a triple interset, then they can always be situated in a hyperplane as in figure 5.25.

Case 2. Six cards left don't form a triple interset.

What about the scenario where we have six cards left at the end of the game, but they do *not* form a triple interset? As we noted in chapter 4, this happens most of the time (roughly 80% of the time that six cards remain, according to simulations in chapter 10). First of all, some of us are a little sad when this happens; finding the pairing that gives a triple interset is "fun." But what is the geometric structure of the six cards pictured in figure 5.26?

It turns out these six cards do not live in any hyperplane—we'll need the entire deck. We can place the cards in the positions occupied by the dots in figure 5.27. You can check that the six cards of figure 5.26

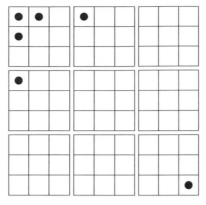

Figure 5.27. Six cards left that don't form a triple interset can always be placed like this.

occupy these positions in figure 5.21. Without getting all technical here, we think the black dots in these positions look a little bit like an arrow.

If you play a game of SET, wind up with six cards at the end, and want something else to do, you should first see if those cards form a triple interset. If they don't, you could take any five of them and place them where the big black dots live in the upper left of the grid in figure 5.27. If you fill out the rest of the grid with cards, completing SETs as you go, then the last card from the six left at the end of the game will wind up in the lower right of the entire configuration.

5.8 REVISITING THE WEIRD FIVE-CARD SCENARIO

In chapter 4, we considered one more situation involving the End Game when six cards are left. Since one card is hidden in the End Game, we have five cards visible in this (fairly common) situation. We asked the following question:

Five-card question: Can any collection of five cards appear?

As we saw in that chapter, the answer to this question is no. It is possible to construct a collection of five cards where the sixth card determined by playing the End Game is one of the five cards we already have. This means that these five cards can't appear at the end of the

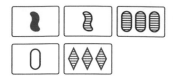

Figure 5.28. Five cards leading to a bad End Game scenario.

game. We now show how to create such five-card collections using the ideas in this chapter.

First, take any four cards that are not in a plane (so the corresponding points are not coplanar). You can ensure your cards satisfy this condition if they don't contain a *SET* and they aren't an interset. For the example, we will use 1 Red Solid Squiggle, 1 Purple Striped Squiggle, 3 Purple Striped Ovals, and 1 Purple Empty Oval.

Next, find the coordinates for the cards, and then find the (mod 3) sum of the cards (recall this procedure from chapter 4). In our case, in order, the coordinates are $(1, 2, 2, 2)$, $(1, 1, 1, 2)$, $(0, 1, 1, 1)$, and $(1, 1, 0, 1)$, so their (mod 3) sum is $(0, 2, 1, 0)$, which corresponds to 3 Red Striped Diamonds, as shown in figure 5.28.

Now, when we work out the missing card using the End Game, we get 3 Red Striped Diamonds, which is already in the layout of cards. (Try this yourself.) This means that these five cards could never have been an actual End Game situation. How did this happen? Recall, the coordinates for 3 Red Striped Diamonds, $(0, 2, 1, 0)$, were the modular sums of the coordinates for the other four cards. That means that the sum of the coordinates of the five cards is $2 \times (0, 2, 1, 0)$, so the only way to make the sum equal to $(0, 0, 0, 0)$ is to add $(0, 2, 1, 0)$ again.

We can now use these cards to produce four more collections of five bad cards. We'll do one, and leave the rest (plus further analysis) to exercise 5.6. We can take four cards from that group of five, but include the 3 Red Striped Diamonds, and now start the process over again with these four cards. We encourage you to do this yourself, using the cards 1 Purple Striped Squiggle, 3 Purple Striped Ovals, 1 Purple Empty Oval, and 3 Red Striped Diamonds. You should find their sum is $(2, 2, 0, 1)$, corresponding to 2 Red Empty Ovals, pictured in figure 5.29.

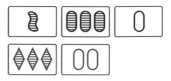

Figure 5.29. Five new cards leading to a bad End Game scenario.

As expected, playing the End Game gives you 2 Red Empty Ovals, the card we added. Here's an interesting thing: We took out 1 Red Solid Squiggle and replaced it with 2 Red Empty Ovals. But these two cards make a *SET* with 3 Red Striped Diamonds, which was the original card we added. Exercise 5.6 will give you a chance to explore this idea.

5.9 CONCLUSION

EUCLID: *I am truly amazed by what happens when we start with a different collection of axioms. I suppose it had never occurred to me that one could get a consistent, abstract geometry by simply changing the axioms. I was just trying to describe the world around me. In my mind, there was only one set of axioms.*

THEANO: *And for thousands of years, it seems that almost everyone thought that way. It took a long time for people to open up to the idea of non-Euclidean geometries. Only within the past two hundred years have these geometries been accepted by the mathematical community. The interesting thing is that, despite their completely abstract conception, some of these geometries do turn out to be useful in the real world.*

SOCRATES: *We have certainly seen how useful affine geometry can be! What other abstract geometries have useful applications?*

THEANO: *All sorts, Socrates! Projective geometry is another abstract geometry that can appear in card games.[20] It is also very*

---

[20] In fact, you can play a version of SET based on projective geometry. For more information, see chapter 9.

important for artists studying perspective. In addition, spherical geometry is important for navigation since the earth is (roughly) spherical, and hyperbolic geometry is used in Einstein's theory of general relativity.

SOCRATES: *I wonder what sorts of geometries will be discovered next!*

THEANO: *Why Socrates, I believe that is the first time in this whole chapter that you have spoken without asking a question.*

SOCRATES: *The question was implied! But I will go ahead and ask it explicitly: What sorts of geometries will be discovered next?*

EUCLID: *That is more of a rhetorical question. Readers, I now address you directly: Go discover some more geometry!*

THEANO: *And when you need a break, play SET.*

EXERCISES

EXERCISE 5.1. Prove directly from the axioms for finite affine geometry that every pair of lines has the same number of points. [Hint: You can do this directly or by contradiction. Directly, you can first show that any two lines that intersect must have the same number of points, and then show that this means that any pair of parallel lines must also have the same number of points. By contradiction, you can assume that you have two lines with different numbers of points on them, and show then that you violate one of the axioms. Note: You'll need to have two cases, since the two lines might intersect, or they might be parallel.]

EXERCISE 5.2. In this exercise, you will show that our four axioms for finite geometry must produce the picture of AG(2, 3) we've come to know and love. We assume that there are three points on each line (see exercise 5.1).

a. Draw three non-collinear points (three points not on the same line), as guaranteed by axiom 1. Label them A, B, and D.
b. Draw the line through A and B, and add a point labeled C to this line.
c. D is not on the line ABC, so by axiom 4, there is a line through it that is parallel to the line you drew. Draw it, and add two more points E and F to that line.

d. You should now have two parallel lines, ABC and DEF. There is a line containing A and D, and it has a third point, G. Which axiom guarantees that this is a new point?

e. There is a line parallel to ADG through B. Why must that line intersect DEF? We can assume that the line contains E, and then add a point H to it.

f. There is also a line parallel to ADG through C. As before, that line must intersect DEF. Show that this line must intersect line DEF at F. Add a point I to this line through C.

g. Show that the line through G and H must contain I.

h. You still need a line through A and E, a line through A and F, a line through B and D, etc. You need every possible line containing a point from ABC and a point from DEF. Draw all those lines, and cite the axioms you used. (When you finish adding all the lines, you should have a nice picture of AG(2, 3), along with an airtight argument for why this picture is "forced.")

Congratulations! You've drawn the unique finite affine plane with three points on a line.

EXERCISE 5.3. When we completed a hyperplane from a plane and one extra card, we ordered the missing 17 cards in figure 5.19

a. Explain why any card in the rightmost plane could be the first card we determine.

b. Explain why there are $9 \times 16!$ possible orders we could have used in completing the hyperplane. (By the way, $9 \times 16! = 188,305,108,992,000$, a very large number.)

EXERCISE 5.4. We can define *parallel classes* of SETs, where two SETs are in the same parallel class if and only if they are parallel to each other. (For those in the know, "parallel" is an equivalence relation, so we're really talking about the equivalence classes of that relation. We'll see this again in exercise 8.3.)

a. Show that there are exactly four parallel classes among the SETs that differ in only one attribute.

b. Explain why any two parallel classes contain the same number of SETs.

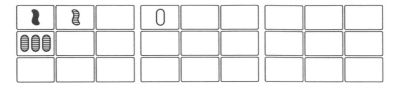

Figure 5.30. Exercise 5.6.

c. How many parallel classes are there among the *SET*s that differ in exactly two attributes? Show your answer is correct twice: first, by appealing to part (b) and then by counting directly.

d. How many parallel classes are there among the *SET*s that differ in exactly three attributes? Show your answer is correct twice: first, by appealing to part (b) and then by counting directly.

e. How many parallel classes are there among the *SET*s that differ in all four attributes? Show your answer is correct twice: first, by appealing to part (b) and then by counting directly.

EXERCISE 5.5. Suppose you've played SET, and you had six cards at the end. Explain why, if there is one interset among the six cards, then the six cards are actually a triple interset.

EXERCISE 5.6. Five bad cards: Recall, in this chapter, we started with four cards that didn't lie in the same plane (so, no *SET*, no interset), 1 Red Solid Squiggle, 1 Purple Striped Squiggle, 3 Purple Striped Ovals, and 1 Purple Empty Oval. We found the coordinates, and added them, mod 3, to get 3 Red Striped Diamonds. We played the End Game with those five cards, and got 3 Red Striped Diamonds. Next, we took out the 1 Red Solid Squiggle from the five cards and did the same thing again. The new card we got, 2 Red Empty Ovals, created another five-bad-cards scenario. We're going to explore this further.

a. First, make a hyperplane starting with the cards shown in figure 5.30. Fill out the hyperplane, and find where 3 Red Striped Diamonds is. Isn't it in a nice spot? Next, we took the four cards that weren't the 1 Red Solid Squiggle, and found that the card that makes a bad End Game scenario is 2 Red Empty Ovals. Find that card in the plane.

b. Use the coordinates of the cards to explain why, when you take four of the five cards, leaving out a card C, the new card that you get must make a *SET* with 3 Red Striped Diamonds.

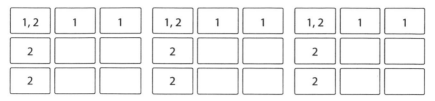

Figure 5.31. Exercise 5.7: The cards with a 1 on them are a hyperplane, and so are the cards with a 2 on them.

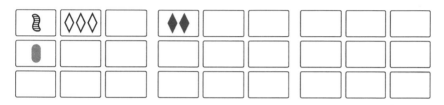

Figure 5.32. Exercise 5.7: Start with four cards, as shown.

c. Use the fact from (b) to find three more bad End Game scenarios (and verify that they work), without using coordinates. Locate the cards in the hyperplane.

EXERCISE 5.7. Making hyperplanes. In this exercise, we'll look at another way to make hyperplanes. Consider figure 5.31.

In the figure, the cards in the position labeled 1 make a subplane. Similarly, the cards in the position labeled 2 make another subplane. (You can verify this with some of the hyperplanes from the chapter, like in figure 5.20.) You can use this to make a hyperplane that starts with four cards, positioned as in figure 5.32.

a. Taking the top rows from each plane and lining them up, you get the start of a plane, as shown in figure 5.33.

Complete this plane, and then put the three rows of cards in the top rows of the hyperplane.

b. Next, take the left columns from each plane, and line them up to get the start of another plane, as shown in figure 5.34.

Complete this plane, and then put the three rows of cards in the left columns of the hyperplane.

c. Now, finish all the planes to see the final hyperplane.

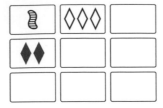

Figure 5.33. Exercise 5.7: One side of our hyperplane, incomplete on the left and complete on the right.

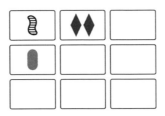

Figure 5.34. Exercise 5.7: Another side of our hyperplane.

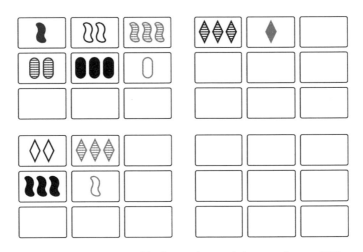

Figure 5.35. Project 5.1: Twelve cards containing exactly two *SET*s.

PROJECTS

PROJECT 5.1. How many *SET*s can there be in a layout of 12 cards? In this project, you'll not only want to find layouts that contain the requested number of *SET*s, you'll also want to show how the *SET*s sit in AG(4, 3). For an example, see the picture in figure 5.35 to see one configuration with two *SET*s in it. You should be able to see from the remaining cards that there won't be any other *SET*s. In what follows, you may use the Cap Builder (described in chapter 9, and available at http://webbox.lafayette.edu/~mcmahone/capbuilder.html).

a. Find a layout of AG(4, 3) with 12 cards that have no *SET*s. The layout should make it clear that there are no *SET*s.
b. Find a layout of AG(4, 3) with 12 cards that has exactly one *SET*. The layout should make it clear that there is only one *SET*.
c. The figure shows one collection of 12 cards with exactly two *SET*s. Now, find a layout of AG(4, 3) that has exactly two *SET*s that intersect each other.
d. There are lots of configurations with exactly three *SET*s. Show one of them in AG(4, 3) so it's clear that there are exactly three *SET*s.
e. The maximum number of *SET*s in 12 cards is 14. Find a configuration of 12 cards that achieves this maximum.
f. Show by example that it is possible to have every possible number of *SET*s in 12 cards, from the minimum of 0 to the maximum of 14. You've done five of them, so you've only got 10 to go!

Interlude: How to Improve at SET

Perhaps you're starting to wonder if all this mathematics is actually going to help you get better at the game. The answer, sadly, is no. While the game happens to be an excellent way to study a variety of branches of mathematics, knowing this will probably not help you find *SET*s faster. But that doesn't mean that there aren't ways to improve! Here are some hints and strategies for improving your game.

I.1 HOW TO FIND *SET*S FASTER

Play lots and lots of SET. Teach the game to your family and friends, and insist that they play with you. Download the SET apps, on your phone and/or tablet. Play online at

- http://www.setgame.com.

This is the company's website. Every day, they put up a new Daily Puzzle, consisting of 12 cards containing six *SET*s, and they time you as you try to find all six. The second of those websites allows you to play a full game, and it also includes some of the mathematics we discuss in this book.

This is really the only way to find *SET*s faster: play *a lot*. If you can, play a lot with someone who's better than you—while it's humbling, you really are forced to get better. As you play more and more, what will eventually happen is that the *SET*s will start to "pop out" at you. In other words, you will get to a point where you don't have to check all the attributes—you'll start seeing *SET*s in their entirety.

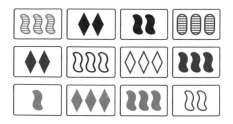

Figure I.1. First layout of cards.

I.2 HOW TO DETERMINE IF THERE ARE NO *SETS* IN A GIVEN CONFIGURATION

In general, the following algorithm will be useful any time you're having trouble finding a *SET*. It allows you to identify every *SET* in a given configuration, so if you follow it carefully and find no *SET*s, that means you need extra cards. It is especially important to be able to do this if you download the SET apps, since they will deduct a point from your score if you mistakenly click the "no *SET*s" icon when there actually is a *SET* in the configuration.

The brute-force way to do this is to check every pair of cards to see if the card that completes the *SET* is there, but this is incredibly slow and inefficient.[1] Our algorithm involves selecting an attribute and mentally dividing the cards into expressions of that attribute.

I.2.1 A Game of SET

Three friends, Satchmo, Ella, and Thelonious,[2] are playing a game of SET. They are about to demonstrate how to use our algorithm. See figure I.1, and follow along as they refer to specific cards.

SATCHMO: *This is a fun game and all, but I don't see any SETs. Do either of you?*
ELLA: *Hey, I see a SET!*
THELONIOUS: *I see several SETs.*
SATCHMO: *Really? How many are there?*
THELONIOUS: *I believe there are four, but we can verify this.*

[1] Unless you are a computer. And we're not computers.
[2] Louis "Satchmo" Armstrong, Ella Fitzgerald, and Thelonious Monk were three jazz greats. While we are pretty sure they never played SET, we can pretend, can't we?

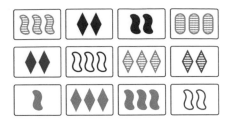

Figure I.2. Second layout of cards.

ELLA: *Yes, we can! It actually won't be too bad—I see there is only one oval. Do you notice that?*

SATCHMO: *Hmm yeah, I do now. Hang on, that's important.... That means that if there is a SET in here, then it'll either be all squiggles, all diamonds, or it'll contain that card!*

THELONIOUS: *Well reasoned. So, let's start by checking the diamonds.*

ELLA: *There are only four diamonds, and there are no SETs among them.*

SATCHMO: *Okay, then I guess we should check the squiggles. There are seven of them. Let's see: Among the solid squiggles, there's a SET. It's all different colors, and numbers.*

ELLA: *That's so. There's also a SET with the only striped squiggle card—it's all 3s. That takes care of the squiggles.*

THELONIOUS: *Now we just need to check to see if there are any SETs with the oval card. If there's a SET with that card, it will necessarily contain all three shapes. That means we actually only have to check it against one of the other expressions: squiggles or diamonds. There are fewer diamonds, so we should check it against those.*

SATCHMO: *All right, I'm pairing diamond cards with the oval card. I see a SET with the 3 Red Empty Diamonds. And another one with the 3 Green Solid Diamonds. That's four SETs already.*

ELLA: *That must be all of them, because we finished this line of reasoning. Everything's copacetic.*

One of the SETs (the all-red SET) is selected and then replaced. See figure I.2.

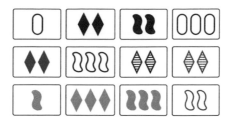

Figure I.3. Third layout of cards.

SATCHMO: *Okay, now that I know what to look for, I notice there is again only one oval—a different one from before.*

ELLA: *Right, but we used shape last time. Let's isolate a different attribute this time. Looking at numbers, I notice there's only one card with 1 symbol.*

THELONIOUS: *Correct, and there are five 2s, and six 3s.*

SATCHMO: *I'll check the 2s. I don't see any SETs in the 2s.*

ELLA: *I don't either. But I'm looking in the 3s, and there is one SET there.*

THELONIOUS: *Precisely. The 3s were easy, because there were no reds, and only one purple, so it had to be all green. It is all the 3 green striped cards.*

SATCHMO: *Now we just have to check cards against the 1 card. Let's check the 2s, since there are fewer of them. I just checked the 1 card against all the 2 cards, and there are no SETs among them.*

ELLA: *I get the same thing. So the only SET is the green one. Let's grab it!*

The green *SET* is taken and then replaced. See figure I.3.

THELONIOUS: *This time, I notice there are only two striped cards.*

SATCHMO: *Ain't no SET there. So let's check the solids and the empties.*

ELLA: *None in the empties and none in the solids.*

SATCHMO: *How'd you do that so fast? There are six solids!*

ELLA: *Yes, but we can apply the same general algorithm at smaller and smaller levels, once we've isolated an attribute. If you take a look at all the solid cards, you'll notice that all the 2 cards are purple or red, and all the 1 and 3 cards are green. So we can't*

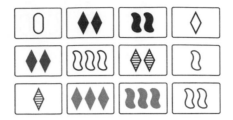

Figure I.4. Fourth layout of cards.

have a 1–2–3 SET in here. If there were a SET in the solids, it'd
have to be the same in number, but we don't have that.

SATCHMO: *I see. Once you get to that point, it's a lot quicker.*

THELONIOUS: *It sure is. Now we just have to check the striped cards
against the empty cards, one at a time. There are no
SETs with the striped purple card.*

ELLA: *I just checked the striped red card, and I found two SETs with
that card.*

SATCHMO: *Those came outta nowhere! How did no one notice them
until now?*

THELONIOUS: *Sometimes when you've been staring at some of the
same cards for too long, you can become SET-blind to
various SETs containing them.*

ELLA: *"SET-blind"?*

THELONIOUS: *It's a term I made up just now, while playing this
game.*

SATCHMO: *I like it!*

ELLA: *Me too. Anyway, let's take one of the SETs.*

One of the SETs is taken and then replaced. See figure I.4.

SATCHMO: *This time, let's look at color. I see only two greens.*

THELONIOUS: *Nice. Obviously there are no SETs in the greens. There
are four reds, and I don't see a SET in those, either.*

ELLA: *Now we have to check the purples. There are more of those, but
if you notice, all the 2-symbol purple cards are solid or striped,
and all the 1 and 3 cards are empty. This means that if there is
a SET in the purple cards, it can't be 1–2–3. So we just have to
check each number.*

SATCHMO: *Yeah, I see what you mean now about using the same process to keep isolating attributes. That really does make it easier, because there aren't even enough 1 or 3 purple cards to make a SET. There are three 2 purple cards, and those don't form a SET. So there are no purple SETs!*

THELONIOUS: *Finally, we have to check for all-different-color SETs. We need to pair each green card up with each of the red cards to see if any of those pairs make a SET.*

ELLA: *I don't find any with the green diamond card.*

SATCHMO: *And I don't find any with the green squiggle card.*

THELONIOUS: *I officially declare this configuration free of SETs.*

SATCHMO: *Well then put out more cards.*

ELLA: *I would, but we're out of cards.*

SATCHMO: *Whew! That game went a lot faster than our previous one, now that we have this algorithm.*

THELONIOUS: *A lot of things in life go faster when you have an algorithm. But you already know this, because you're a musician! It's like I always say: "All musicians are subconsciously mathematicians."*

ELLA: *He really did say that, by the way. Look it up!*

SATCHMO: *Hey cool! That means I'm a mathematician, and so are you!*[3]

Indeed, they are correct. The configuration in figure I.4 is an example of 12 cards that do not contain a *SET*. (And they are also correct that musicians are subconsciously mathematicians, but don't worry, you can still be a mathematician even if you are not a musician.)

The systematic thought process they went through to identify all the *SET*s in each configuration is an algorithm you can use any time you're playing the game, and it is considerably faster than checking every pair of cards. It is especially useful in configurations where one expression of an attribute is missing (or nearly missing): any time you happen to notice, for instance, that there are no red cards, or just one squiggle, or only two 3s, the game suddenly gets a lot easier.

---

[3] Here, Louis Armstrong could be referring to either of his fellow musicians, or he could be referring to *you*, the reader. This is open to interpretation.

You should find this process most useful if you've been looking for *SET*s for a while and get to a point where you want to be sure there are no *SET*s. As you continue to play SET, keep an eye out for missing expressions of an attribute, or expressions with only one or two cards— noticing this will help you narrow down possible *SET*s a lot faster.

The Algorithm

We wrap up this procedure in a tidy, step-by-step package:

1. Select an attribute (number, color, shading, or shape). Preferably, you will select an attribute for which one of the expressions is underrepresented.
2. Mentally (or physically, if this is possible) divide the cards into the different expressions of that attribute. For instance, if you selected color, make three groups: put the red cards together, the green cards together, and the purple cards together.
3. Check each of the three groups individually for *SET*s.[4] This will allow you to find every *SET* where the attribute you selected is all the same.
4. Finally, identify whichever group has the least number of cards, and check each card in that group against each card in the group with the second least number of cards. This is the fastest way to find *SET*s where the attribute you selected is all different.
5. Didn't find any *SET*s? Now you may confidently put out three more cards, knowing that if there is a *SET* in your new configuration of 15 cards, it must include at least one of the cards you just put out.

I.3 FAIR(ER) GAMES BETWEEN TWO PLAYERS OF DIFFERENT SKILL LEVELS

We've played a lot of SET. We're pretty good at it. So, how do we play with novice players and still keep the game interesting? Our friend

---

[4] At this point, if one of your groups has six or more cards, you can actually start the algorithm over with a different attribute for that group of cards (the way Ella did), as this may speed up the process.

Carolyn Chun[5] sent us suggestions for ways to play, and we've added some ourselves.

1. After finding a *SET*, count to 10 (or 20, or 30, ... ; experiment to see what works best) before you are allowed to take it. You may do this out loud, so others become aware that there is a *SET* on the table, or you can do it in your head if they find it distracting.

2. You must find two disjoint (non-intersecting) *SET*s before you are allowed to take one. This slows you down, and it also means that every time you take a *SET*, others will know there is another *SET* on the table. (You can also try this variation without the disjoint condition, which will make it easier, but it will also mean you are potentially depriving others of *SET*s.)

3. Novice players get to pick a card from the opening configuration. The expert player cannot take a *SET* that uses any card that has been selected. Novice players get a bonus point when they find a *SET* with their card, and they also get to replace it with a new card. (This is a variation of "Pick a Card," which is item (6) from the next section on alternate ways to play.)

4. Look at the cards from a different angle from what you are used to, and/or sit farther away from the cards than the other players.

5. Make up your own handicap. Maybe you have to sing a song, or do a dance, or do something else silly every time you find a *SET* before you are allowed to take it. Maybe you are allowed to use the green cards only. Maybe you aren't allowed to use the word "*SET*!" There are a lot of ways to play slower to give novice players the chance to catch up to you.

I.4 HOW TO USE THE CARDS IN OTHER WAYS

For as long as SET has been around, people have been inventing alternate games to play with the cards. They can be a nice break from regular SET, and some will actually make you a better player.

---

[5] She's very good at SET, too.

1. **The End Game.** We introduced the End Game in chapter 1 and explained why it works in chapter 4. As you may recall, we can hide a card at the beginning of a game of SET, then determine that missing card from the cards showing at the end of the game. The great thing about the End Game is that it comes at the end of a regular game, so you can do it every time you play a game of SET. What's really fun is looking for *SET*s with the missing card before it has been revealed. Since this is a lot harder than looking for *SET*s with cards you can actually see, it will make the regular game seem easier by comparison. In fact, we no longer keep track of how many *SET*s each individual person took. We just play to get to the End Game.[6] There's even an online version now, available at http://www.bluffton.edu/homepages/facstaff/nesterd/java/setendgame.html.

2. **Interset.** If you're feeling adventurous, deal the cards as usual, but look for intersets instead of *SET*s. As we learned in chapter 2, the expected value (average number) of intersets in a layout of 12 cards is about 19.[7] So, unlike in the regular game, you will not have to worry about potentially needing more cards— you should expect to find *a lot* of intersets. (This isn't to say that they are easy to find. In fact, because there's no real pattern recognition, they can be a lot harder to find than regular *SET*s.) You can play "Interset" as you would play the regular game, by clearing intersets and replacing them, or you can just try to see how many you can find in a given layout of 12 cards in a specified amount of time. Bonus points for triple intersets (or higher).

3. **SET–Planet–Comet.** If you're feeling even more adventurous, you can play a version of the game invented by a group of mathematicians in which nine cards are dealt and players look for either *SET*s, planets, or comets.[8] They defined a planet as four cards in a plane (so either an interset or a *SET* plus one card, but the interset option is more interesting, since the *SET*-plus-a-card

---

[6] Partly, this is because we're pretty sure we know who won.

[7] Actually, we learned it was about 18.8, but since we don't know what 0.8 of an interset is, we'll just round up.

[8] M. Baker et al., "Sets, planets, and comets," *College Mathematics Journal* **44**, no. 4 (September 2013), 258–264. See question 8 in chapter 3 for more information about this paper.

option would be more likely to be taken just as a *SET*), and they defined a comet as nine cards whose coordinate-wise sum is 0, i.e., nine cards that can be split up into single-attribute *SET*s. (Since their work was independent of our work, the terminology is different, as often happens in mathematics.) The neat thing about this version of the game is that you will never need to add more cards, as it was proved that any collection of nine cards must contain a *SET*, a planet, or a comet. However, since you'll be taking things that aren't *SET*s, you won't be able to play the End Game afterward.

4. **Clear the Deck.** One way to do this is to play a regular game of SET and have no cards left, but this almost never happens. If you've played a regular game and have cards left, one thing you can do is take your pile of *SET*s and try to reorganize *SET*s until you've used all the cards. If you're playing with others, you can split up the pile of *SET*s, and each person can take a turn choosing a *SET*, putting it down next to the leftover cards, and looking for different *SET*s. If no new *SET*s are found, the *SET* should be removed and replaced with a different one, but if new *SET*s are found, those should be taken instead. This can go on for a while, but if you look carefully, you may eventually notice a configuration of *SET*s that would allow the deck to be cleared.

5. **Five-Attribute SET.** As we've mentioned, it is possible to obtain three decks of SET cards, draw something in the background of all of the cards in one of the decks, and draw something different in the background of all the cards in another one of the decks. When we did this, we drew thick diagonal lines going across the cards (without covering the attributes—we made it look as though the lines went "under" the attributes) on one deck, and thin diagonal lines going across in the other direction on another deck. One deck is left unadulterated. Combining all three of these decks allows you to play Five-Attribute SET.[9] Play as long as you can before you get a headache, and then, go back to playing regular SET—you'll be amazed how much easier it seems by comparison.

---

[9] We've heard this called Evil SET. The name is appropriate.

6. **Pick a card. Any card?** Each player gets to choose one of the cards on the table. Players can't take any *SET*s that use someone else's card, and they get a bonus point when they find a *SET* with their own card (at which time they would select a new card). This can encourage players to develop an intuition for which cards in a given layout are most likely to be in *SET*s. (Thanks, Carolyn Chun!)

7. **SET Solitaire.** Of course, you can play SET in the usual way by yourself, but it's fun to introduce variations. We learned this variation from Alexa Kottmeyer.[10] Lay out the cards in three separate groups of nine cards. You may take only *SET*s that have one card in each pile.

8. **Take a *SET*, leave a *SET*, Percocet.** This version of the game doesn't exist—yet. It's a catchy title, but we haven't actually thought of any rules to go along with the title that would make sense. Can you?

9. The company that makes SET also has some alternate versions of the game on their website http://www.setgame.com/teachers corner/other-ways-to-play. They include our End Game, which is pretty cool.

10. Make up your own version of the game. The possibilities are endless!

I.5 HOW TO MAKE OTHERS PLAY WORSE

Finally, we offer (somewhat facetiously) a last resort method to winning: preventing others from winning. There are many ways to do this.

1. While you're all sitting around the cards, start hovering over whichever cards are closest to you. This will have the effect of both intimidating others as well as preventing them from seeing certain cards.

2. This method will work only if you're the dealer. When you deal the cards, you can lay them out very slowly and look at them

---

[10] You won't be surprised to learn she's very good, too.

before you place them out for everyone else to see. If you can see the cards first, then you can find the *SET*s first.

3. Pretend you've found a *SET* when you haven't. You have to be careful with this one, since some people will insist on deducting a point if you call "*SET!*" and turn out to be wrong, so you'll have to make sure not to actually call "*SET!*"—just suddenly start reaching toward some of the cards, then quickly change your mind. This will startle/annoy other players, but you can only get away with doing it once or twice, since they'll quickly become desensitized to your feints.

4. Wait for someone else to call "*SET!*" The instant they begin, yell "*SET!*" louder than them. Often, people assume whoever said it louder must have said it first. With any luck, the person who actually found a *SET* will have begun to take the cards, or at least have indicated roughly where the *SET* is, so that you can easily steal it. If they haven't, though, you'll have maybe three seconds of frantic searching to find it before people will start doubting that you actually found one.[11]

5. Make obnoxious noises and/or sing obnoxious songs. This is both fun for you and incredibly annoying for others, who will find it distracting.

6. Distract the people around you by pointing out interesting objects in the room. When they turn to look at whatever you're pointing at, steal a *SET* from their pile of cards and say, "Oh, I guess it was nothing. And I definitely didn't steal one of your *SET*s."

7. Have a cat around when you play, as in figure I.5. Nothing attracts a house cat like a group of people sitting around some cards on the floor. What will most likely happen is your cat, noticing where everyone's attention is concentrated, will walk right on top of the cards and momentarily bask in everyone's attention (until someone "encourages" the cat to move). This will prevent you from seeing some of the cards, but the good news is that it will prevent others from seeing those cards as well.

---

[11] This technique is probably the one that is most likely to start fights.

Figure I.5. Cats looking for *SET*s.

8. Finally, if all of these "methods" fail—and they will, as soon as your fellow players decide not to tolerate your ridiculous antics any longer—turn on the TV. We have found that television is an excellent way to stop all original thought. Then, while everyone is watching TV, find all the *SET*s, or just steal everyone else's *SET*s. Congratulations, you are now the SET champion.

As you can see, there are a variety of ways you can improve your game. Mostly, though, the best advice is to go forth and play SET! And when you feel like taking a break, read more about the mathematics underlying the game. We recommend this book.

What's up next? The remaining chapters of this book will explore some of these mathematical concepts in more depth, and at a higher level. Don't let that discourage you! You can take as much time as you want with some of the sections, and/or skip some of them. After all, this is your book.

More Combinatorics

6.1 PREAMBLE

The first five chapters of this book introduced you to many of the connections between mathematics and SET. We had a large cast of characters to help explain key ideas. We leave these folks at this point in the book, presumably to go play SET. We hope the reward for working through the details of what follows will be an enhanced understanding of some deeper topics.

6.2 INTRODUCTION

In chapter 2, we saw that SET gives us lots of things to count. In this chapter, we turn to some more advanced counting questions, usually in higher dimensions. We will often count the same thing in two different contexts: *global* counts, where we count a total for the entire deck, and *local* counts, where our count is tied to one fixed, specified object (a card, a *SET*, a plane, and so on).

We start with a global count. SET involves four attributes (number, color, shading, and shape), each of which can have three different values. One way mathematicians often generalize the game is to add attributes, while insisting that each attribute still has only three possible values. This means that a *SET* will always consist of three cards, and two cards will always uniquely determine a *SET*.

For example, suppose we wish to play seven-attribute SET. Then, in addition to number, color, shading, and shape, we could add

- flavor: each card is chocolate, vanilla, or strawberry;
- category: each card is animal, vegetable, or mineral;
- size: each card is small, medium, or large.

For instance, let's choose two cards in our seven-attribute game: 2 Red Empty Large Animal Chocolate Squiggles and 2 Green Empty Small Vegetable Vanilla Squiggles. Then, as in the usual game, there is a unique card that completes a *SET* (2 Purple Empty Medium Mineral Strawberry Squiggles). We conclude that the fundamental theorem of SET remains true in the higher-attribute game.

We can (and will) ignore the specific attributes when doing our computations. The advantage of deriving formulas that depend on n is obvious. Having one formula to rule them all[1] allows us to answer an infinite number of questions. Here are a few questions we'll consider:

- How many cards are in a complete deck?
- How many *SET*s are there? How many *SET*s contain a given card?
- How many *SET*s have no attributes in common? How many have exactly one attribute in common? More generally, how many have k attributes in common, where $0 \leq k \leq n - 1$?
- How many planes, hyperplanes, and higher-dimensional planes are there?

The basic structure of our questions is the following:

Let a_n denote the number of _____ (where we fill in the blank with "cards," "*SET*s," "*SET*s with a given number of attributes the same,"...) in the n-dimensional version of SET. Find a formula for a_n in terms of n.

We answer the first two questions now, mostly as a warm-up. How many cards are in a complete deck for the n-attribute game? Since we have three choices for each attribute, there are 3^n cards.

- There are 3^n cards in a complete deck.

When $n = 4$, this gives us the right answer, $3^4 = 81$ cards.[2]

---

[1] This is a reference to something, we think.

[2] Well, it better. It's a good idea to check your formulas, assuming you know the answers for some small cases. If we got the wrong answer, it would probably be a good idea to revise the formula.

TABLE 6.1.
The number of *SET*s in n-attribute SET, for $n \le 7$.

# attributes	1	2	3	4	5	6	7
# *SET*s	1	12	117	1080	9801	88,452	796,797

How many *SET*s are there in the n-attribute game? To answer this, just note that, by the fundamental theorem, it is still true that two cards determine a unique *SET*. As in chapter 2, we have $3^n(3^n - 1)/2$ ways to choose two cards from the deck of 3^n cards. But this counts a given *SET* three times since there are three ways to select two cards to produce the same *SET* (again, exactly as in chapter 2). This gives us a total of $3^n(3^n - 1)/6$.

Simplifying gives us our formula in terms of n:

• The total number of *SET*s in n-attribute SET is $3^{n-1}(3^n - 1)/2$.

We give the total number of *SET*s for some small values of n in table 6.1.

Let's check that this formula agrees with what we already know. There are 12 *SET*s in a plane (the two-attribute game), 117 *SET*s in a hyperplane (the three-attribute game), and 1080 *SET*s in the usual, four-attribute version. We'll generalize this formula in section 6.5 by finding the number of planes, hyperplanes, and so on, in n-attribute SET.

6.3 LOCAL COUNTS

In this section, we consider two local counts for the n-attribute game:

• How many *SET*s contain a given card?
• How many (other) *SET*s intersect a given *SET*?

The argument we use to answer the first question is essentially the same as the one we gave in chapter 2. First, choose a card C from the n-attribute deck. We can break up the $3^n - 1$ remaining cards in the deck into $(3^n - 1)/2$ pairs, where each pair forms a *SET* with our chosen card C. This depends only on the fundamental theorem, so we have an answer:

• Each card is contained in $(3^n - 1)/2$ *SET*s.

TABLE 6.2.
The number of *SET*s meeting a given *SET S* in *n*-attribute SET, for $n \leq 7$.

n	1	2	3	4	5	6	7
# *SET*s meeting S	0	9	36	117	360	1089	3276
Percentage of total	0%	75%	30.8%	10.8%	3.7%	1.2%	0.4%

For the second question, we note that counting the number of *SET*s meeting a given *SET* is important when you play the game. For instance, if two *SET*s have a card in common, then removing one *SET* destroys the other *SET*.

Here's a recipe for counting the number of (other) *SET*s that intersect a given *SET* in our *n*-attribute game.

1. First, choose your *SET*. Call the three cards A, B, and C.
2. How many *SET*s use card A? By the computation we just did, there are $(3^n - 1)/2$ such *SET*s. Note that this includes our *SET* ABC.
3. Now repeat this for cards B and C. Adding these results (temporarily) gives us $3 \times (3^n - 1)/2$.
4. But the answer from part (3) includes our original *SET* ABC three times, so we need to subtract 3. This gives us a total of $3 \times (3^n - 1)/2 - 3$.

After simplifying, we conclude the following:

- The number of other *SET*s that meet a given *SET* is $\frac{3}{2}(3^n - 3)$.

It can be fun[3] to look at some data for the number of *SET*s intersecting a given *SET*. In table 6.2, we give the number of *SET*s that meet a given *SET* and the percentage of all *SET*s this represents.

There is one number that we would like to focus on: the 117 *SET*s that intersect a *SET* in the four-attribute game. We've seen the number 117 before: it's the *total* number of *SET*s in the three-attribute game (from table 6.1)! Is this a coincidence?[4]

The answer, unfortunately,[5] is yes. It can be hard to convince yourself that something is genuinely a coincidence, but it's often worth

---

[3] To be fair, this depends on your definition of "fun." Greatly.
[4] Whenever you see this question in a math book, the answer is always no. Except this time.
[5] Maybe we should be happy. We're not sure.

TABLE 6.3.
Comparing the total number of *SET*s and the number of *SET*s meeting a given *SET S*.

n	1	2	3	4	5	6	7
# *SET*s	1	12	117	1080	9801	88,452	796,797
# *SET*s meeting S	0	9	36	117	360	1089	3276

TABLE 6.4.
Number of *SET*s with 0, 1, 2, or 3 attributes the same.

# attributes the same	# SETs	Percentage
0	216	20%
1	432	40%
2	324	30%
3	108	10%
Total	1080	100%

the effort. Here's one explanation. If there were a theoretical reason these numbers should be the same, we should see the same relationship hold when the number of attributes is not four.

How can we check if there is a connection between the number of *SET*s in $(n-1)$-attribute SET and the number of *SET*s that intersect a given *SET* in n-attribute SET? In table 6.3, we compare the information in tables 6.1 and 6.2 for $n \leq 7$.

Moral: Sometimes two counts give the same answer by accident.

6.4 COUNTING THE NUMBER OF *SETS* WITH k ATTRIBUTES IN COMMON

In chapter 2, we partitioned the deck of cards into four classes, depending on the number of attributes that were the same. We summarize what we found in table 6.4.

Our goal in this section is to figure out what happens in n-attribute SET when $n \neq 4$. We'll do two versions of this problem: a global count for the whole deck and a local count for the number containing a given card. It should be obvious that the local answer won't depend on the card we've picked. Calculating the number of *SET*s with exactly

k attributes the same will involve several standard techniques used in combinatorics. We'll need the *binomial coefficients*, which we met in chapter 2:

- Let $S = \{1, 2, \ldots, n\}$; the number of subsets of S of size k is

$$\binom{n}{k} = \frac{n!}{k!(n-k)!}.$$

6.4.1 Global Counts

We begin by introducing some notation. Let $g(n, k)$ equal the number of *SET*s in the n-dimensional game with exactly k attributes the same. Then $g(n, k)$ is a function that depends on the two integers n and k, where we assume $0 \le k \le n-1$. So, for instance, $g(4, 1) = 432$ encodes the fact that there are 432 *SET*s in the usual four-attribute game with one attribute the same.

Our immediate goal is to find a formula for $g(n, k)$. To do this, it will be useful to work with the cards as vectors. Recall that a card is represented by an ordered n-tuple $(x_1, x_2, \ldots, x_n)$, where each $x_i = 0$, 1, or 2. Our *SET* will be three cards

$$(a_1, a_2, \ldots, a_n), \qquad (b_1, b_2, \ldots, b_n), \qquad (c_1, c_2, \ldots, c_n),$$

where

- $a_i = b_i = c_i$ for the k attributes that are the same, and
- a_i, b_i, and c_i are all different for the $n - k$ attributes that are all different.

To compute $g(n, k)$, we first choose the k attributes that will be the same. There are $\binom{n}{k} = \frac{n!}{k!(n-k)!}$ ways to choose these k attributes. For instance, if $n = 9$ and $k = 4$, we could select attributes 1, 3, 6, and 8 to be the same. We indicate this by boxing those attributes that are the same in the vectors that represent our cards:

$$\left(\boxed{*}, *, \boxed{*}, *, *, \boxed{*}, *, \boxed{*}, * \right).$$

Now for each of those k attributes, there are three choices for its value. This means that there are $\binom{n}{k} \times 3^k$ ways to specify the attributes that are the same for the *SET*. Returning to our $n = 9$, $k = 4$ example, we make the following arbitrary selections for our boxed attributes:

$$\left(\boxed{0}, *, \boxed{2}, *, *, \boxed{1}, *, \boxed{1}, *\right).$$

Finally, we need to determine the remaining $n - k$ places where the attributes are different. If the three cards in the *SET* are ordered as card 1, card 2, and card 3, then for each of the $n - k$ attributes that are all different, we have three choices for that attribute in card 1, two choices in card 2, and one choice in card 3. This gives $3! = 6$ choices for each of the $n - k$ attributes, giving 6^{n-k} ways to complete the *SET*.

But this overcounts $g(n, k)$ by a factor of 6 because there are $3! = 6$ ways we could have labeled the cards as card 1, 2, and 3. So we have 6^{n-k-1} ways to complete our *SET*. Putting all of this together gives us the answer: $g(n, k) = \binom{n}{k}3^k 6^{n-k-1}$. A little algebra allows us to rewrite this.

- The number of *SET*s with k attributes the same is

$$g(n, k) = \binom{n}{k} 3^{n-1} 2^{n-k-1}.$$

Plugging $n = 9$ and $k = 4$ into this formula gives $g(9, 4) = 13,226,$ 976 *SET*s in the nine-attribute game that have exactly four attributes the same. That's a lot of *SET*s.[6] Here is one such *SET*:

$$\text{card } 1 = \left(\boxed{0}, 0, \boxed{2}, 2, 2, \boxed{1}, 1, \boxed{1}, 0\right),$$

$$\text{card } 2 = \left(\boxed{0}, 1, \boxed{2}, 0, 1, \boxed{1}, 2, \boxed{1}, 1\right),$$

$$\text{card } 3 = \left(\boxed{0}, 2, \boxed{2}, 1, 0, \boxed{1}, 0, \boxed{1}, 2\right).$$

Table 6.5 gives the number of *SET*s with k attributes the same (for $0 \leq k \leq n - 1$) for $n \leq 5$.

---

[6] If nothing else, this should indicate that we have left the world of game playing far, far behind.

TABLE 6.5.

The number of *SET*s $g(n, k)$ in the n-attribute game with exactly k attributes the same.

	$k = 0$	$k = 1$	$k = 2$	$k = 3$	$k = 4$	Total
$n = 1$	1	—	—	—	—	1
$n = 2$	6	6	—	—	—	12
$n = 3$	36	54	27	—	—	117
$n = 4$	216	432	324	108	—	1080
$n = 5$	1296	3240	3240	1620	405	9801

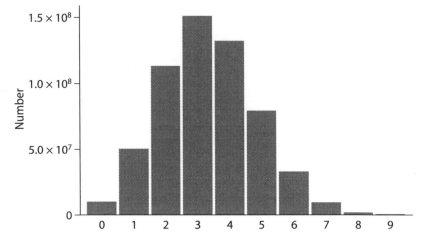

Figure 6.1. The number of *SET*s with k attributes the same (for $0 \leq k \leq 9$) in the 10-attribute game.

It's always nice to visualize data, when possible. We give the distribution of the number of *SET*s with k attributes the same for the 10-attribute game in figure 6.1.

This graph should motivate quite a few questions for the n-attribute game:

- What kind of *SET* is most common, i.e., for what value of k is $g(n, k)$ maximized?
- What kind of *SET* is least common, i.e., for what value of k is $g(n, k)$ minimized?
- What is the *average* number of attributes a *SET* has in common?
- Do the numbers always increase to a maximum, then decrease?
- In table 6.5, note that $g(5, 2) = g(5, 1) = 3240$. What other values of n have ties for the largest value of $g(n, k)$?

We will answer all of these questions, and more, in chapter 7, where we consider the data from a probabilistic point of view.

6.4.2 Local Counts

Pick a card in the n-attribute game. There are $(3^n - 1)/2$ *SET*s that contain that card. How many of those *SET*s have exactly k attributes the same? Let's call this number $l(n, k)$, which, like $g(n, k)$, is a function that depends on both n and k.

To find a formula for $l(n, k)$, we'll use an incidence count, as we did back in chapter 2. To do this, we construct a bipartite graph with all 3^n cards on the left-hand side and the $g(n, k)$ *SET*s with exactly k attributes the same on the right-hand side.

Then each of the 3^n cards is joined to $l(n, k)$ *SET*s, so the total number of edges in the bipartite graph is $3^n \times l(n, k)$. On the other hand, each of the $g(n, k)$ *SET*s on the right-hand side is joined to exactly 3 cards on the left (since each *SET* has 3 cards), giving a total of $3 \times g(n, k)$ edges in the bipartite graph.

Equating these two values for the total number of edges gives us a formula relating the local and global counts:

$$3^n l(n, k) = 3g(n, k).$$

Then we can use our formula for $g(n, k)$ from section 6.4.1 to give us the formula for $l(n, k)$:

- The number of *SET*s with k attributes the same containing a given card is

$$l(n, k) = \binom{n}{k} 2^{n-k-1}.$$

Table 6.6 gives $l(n, k)$ for some small values of n. For instance, we see that $l(4, 0) = 16$, which tells us that 16 (of the 40) *SET*s that contain a given card will have no attributes the same.

Let's reconnect to SET. Choose your favorite card in the deck, 2 Purple Striped Squiggles,[7] and look at the *SET*s containing it. Figure 6.2

---

[7] If this isn't your favorite card, you should write your own book. We'll buy it.

TABLE 6.6.
The number of SETs with k attributes the same containing a given card.

	$k = 0$	$k = 1$	$k = 2$	$k = 3$	$k = 4$	Total
$n = 1$	1	—	—	—	—	1
$n = 2$	2	2	—	—	—	4
$n = 3$	4	6	3	—	—	13
$n = 4$	8	16	12	4	—	40
$n = 5$	16	40	40	20	5	121

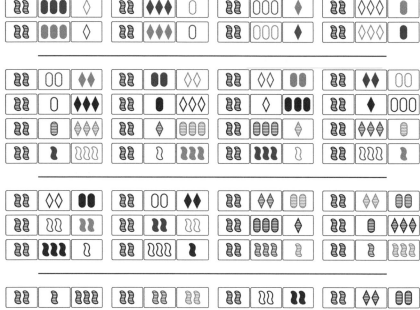

Figure 6.2. All of the SETs containing 2 Purple Striped Squiggles, organized by the number of attributes that are the same.

shows the 40 SETs that contain this card, partitioned into four classes based on the number of attributes that are the same.

Finally, recall that 20% of the SETs in the whole deck have no attributes the same, 40% have one attribute the same, 30% have two attributes the same, and 10% have three attributes the same. The percentages for our local counts are the same: given a specific card, 20% of the 40 SETs that contain that card have no attributes the same, 40% have one attribute the same, 30% have two attributes the same, and 10% have three attributes the same.

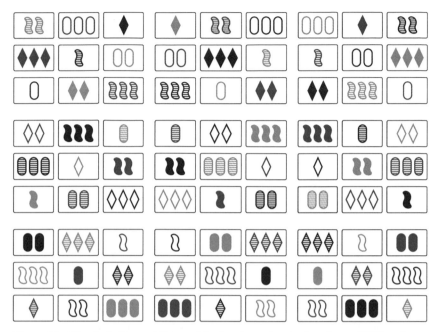

Figure 6.3. The deck. Stare at this and find patterns for the locations of *SET*s, planes, and hyperplanes.

Is this a coincidence?[8] You now have all the ingredients needed to calculate both the local and the global percentages for the n-attribute version of the game—see exercise 6.2 for the details.

6.5 COUNTING PLANES, HYPERPLANES, AND q-BINOMIALS

6.5.1 Make Yourself a Pretty Picture

In this section, we explore some counting questions associated with the geometric structure of the deck. In figure 6.3, there's a beautiful picture of the entire deck similar to one we gave in section 5.5. This figure allows you to "see" all 1080 *SET*s at once.

- How many different ways could we construct such a wonderful picture?

[8] This time, no.

Our procedure for making this picture gives us the answer. We'll use ordered pairs of the form (row, column) to label the positions of the cards, with (1, 1) corresponding to the upper left corner and (9, 9) corresponding to the lower right.

a. Place a card in position (1, 1). There are $81 = 3^4$ choices for this card.

b. Place a card in position (1, 2) immediately to the right of the card you just placed. There are $80 = 3^4 - 1$ choices for this card. These first two cards also determine the card in position (1, 3), because the three cards in positions (1, 1), (1, 2), and (1, 3) form a *SET*.

c. Place a card in position (2, 1). There are $78 = 3^4 - 3$ choices for this card. As we've seen before, the cards in positions (1, 1), (1, 2), and (2, 1) uniquely determine the whole plane in the upper left corner of the grid, i.e., all cards occupying positions (x, y) with $1 \leq x, y \leq 3$ are now fixed.

d. Now choose a card to place in position (1, 4). There are $72 = 3^4 - 3^2$ choices for this card. This card, together with the nine cards already determined, uniquely determines the hyperplane occupying the first three rows of the grid, i.e., all cards in positions (x, y) where $1 \leq x \leq 3$. (This is identical to the procedure we used in figure 5.17.)

e. Finally, choose a card to place in position (4, 1). There are $54 = 3^4 - 3^3$ choices for this card. At this point, the positions of all the remaining cards are completely determined, as we saw in chapter 5.

Putting this all together gives us the number of "different" pictures we can create:

$$3^4(3^4 - 1)(3^4 - 3)(3^4 - 3^2)(3^4 - 3^3) = 1,965,150,720.$$

That's a big number.[9] On the other hand, if everyone on the planet made their own picture, there would be some repeats.[10] More to the

---

[9] For instance, if you and your friend Elvis each create your own pictures of the entire deck, you'll probably wind up with different configurations.

[10] This follows from the pigeonhole principle, and knowing something about the population of the earth. And a vivid imagination.

point, the product $3^n(3^n-1)(3^n-3^2)\cdots(3^n-3^m)$ will be very important when we try to count planes and hyperplanes when $n > 4$, which we turn to now.

6.5.2 Counting Planes and Hyperplanes

From the geometric point of view, a *SET* is simply a line, i.e., a one-dimensional object. We play ordinary SET by removing one-dimensional objects from the collection of cards facing up on the table. In theory, we could play a game where the goal would be to remove a two-dimensional plane, or a three-dimensional (or higher) hyperplane. The point here is that *SET*s are just special cases of a more general structure.

Our first goal is finding a formula for the number of k-dimensional hyperplanes in the n-attribute game, where $1 \leq k \leq n-1$. When $k = 1$, a k-dimensional hyperplane is simply a *SET*, and we've already counted the number of *SET*s in the n-attribute game. So, let's turn our attention to $k = 2$, and count the two-dimensional planes in n-attribute SET. Here's a guide to counting the number of planes.

COUNTING PLANES

- We assume $n \geq 2$. A plane is determined by three non-collinear points. There are 3^n choices for the first point, $3^n - 1$ choices for the second point, and $3^n - 3$ choices for the third (since the third can't be on the line determined by the first two points). This gives us the product $3^n(3^n - 1)(3^n - 3)$.
- However, this procedure overcounts the number of planes. Since the order of the cards within a plane doesn't matter, we need to divide by the number of ways the nine cards in our plane could have been placed: We had $9 (= 3^2)$ choices for the first point we could have chosen, and $8 (= 3^2 - 1)$ for the second, and $6 (= 3^2 - 3)$ for the third. This tells us each plane has been counted $9 \times 8 \times 6 = 432$ times by the above procedure.
- Thus, the number of planes is

$$\frac{3^n(3^n - 1)(3^n - 3)}{3^2(3^2 - 1)(3^2 - 3)}.$$

TABLE 6.7.
The number of k-dimensional hyperplanes in the n-attribute game ($n, k \leq 7$).

	SETs						
	$k = 1$	$k = 2$	$k = 3$	$k = 4$	$k = 5$	$k = 6$	$k = 7$
$n = 2$	12	1	—	—	—	—	—
$n = 3$	117	39	1	—	—	—	—
$n = 4$	1080	1170	120	1	—	—	—
$n = 5$	9801	32,670	10,890	363	1	—	—
$n = 6$	88,452	891,891	914,760	99,099	1092	1	—
$n = 7$	796,797	24,169,509	74,987,451	24,995,817	895,167	3279	1

As a quick check, we plug $n = 2$, 3, and 4 into this formula. This will tell us how many planes are contained in a plane, a hyperplane, and the SET deck, respectively.

When $n = 2$, we get $(9 \times 8 \times 6)/(9 \times 8 \times 6) = 1$: there is one plane contained in a plane. For $n = 3$, we have $(27 \times 26 \times 24)/(9 \times 8 \times 6) = 39$ planes, which agrees with our calculation from chapter 5. For the usual four-attribute SET, we get $(81 \times 80 \times 78)/(9 \times 8 \times 6) = 1170$ planes, which agrees with what we discovered in chapter 2.

The procedure we used to count the number of planes suggests a general formula. We let $h(n, k)$ represent the number of k-dimensional hyperplanes for $k \leq n$. Then

$$h(n, k) = \frac{3^n(3^n - 1)(3^n - 3)(3^n - 3^2) \cdots (3^n - 3^{k-1})}{3^k(3^k - 1)(3^k - 3)(3^k - 3^2) \cdots (3^k - 3^{k-1})}.$$

Table 6.7 contains several values of $h(n, k)$.

Several comments are in order:

1. First, it's rather amazing that this fraction is always an integer, i.e., all the stuff in the denominator cancels out. This should remind you of what happens for the binomial coefficient $\binom{n}{k}$, where the denominator $k!(n - k)!$ is completely canceled out by factors of the numerator $n!$.

2. The value of $h(n, 1)$ is the number of one-dimensional hyperplanes, i.e., the number of SETs. Indeed,

$$\frac{3^n(3^n - 1)}{3(3 - 1)} = \frac{3^{n-1}(3^n - 1)}{2},$$

which agrees with the formula we found in the introduction to this chapter.

3. When $n = k$, the formula reduces to 1, i.e., there is one hyperplane of dimension n. This is a trivial case, of course, but it's nice that the formula gives us something reasonable.

4. Finally, an examination of table 6.7 may convince you that the numbers in any row seem to increase to a maximum, then decrease. A sequence of numbers with this property is said to be *unimodal*. There are lots of interesting sequences that are unimodal, and there are some very challenging open problems in combinatorics concerning unimodal sequences.

6.5.3 How Many Hyperplanes Contain a Given Card?

We conclude this section with a local version: How many hyperplanes of dimension k contain a given card? We've done one version of this problem already:

• When $k = 1$ and n is arbitrary, we found each card is in $(3^n - 1)/2$ *SET*s.

To find a general formula, we can use another incidence count. Here's the setup: There are 3^n cards on one side, $h(n, k)$ hyperplanes of dimension k on the other side. As usual, we join a card to a hyperplane of dimension k if the card is in the hyperplane. We know that each such hyperplane contains 3^k cards. We let x represent the unknown number of hyperplanes containing a given card. Then we get $3^n x = 3^k h(n, k)$, so

$$x = \frac{h(n, k)}{3^{n-k}}.$$

Then, using our formula for $h(n, k)$ and a bit of cancelation, we now have a formula for x:

$$x = \frac{(3^n - 1)(3^n - 3)(3^n - 3^2) \cdots (3^n - 3^{k-1})}{(3^k - 1)(3^k - 3)(3^k - 3^2) \cdots (3^k - 3^{k-1})}.$$

It turns out these numbers are famous, and there is standard notation for them. The number of hyperplanes of dimension k containing a given card is written $\begin{bmatrix} n \\ k \end{bmatrix}_3$. These are the *q-binomial*

TABLE 6.8.

$\begin{bmatrix} n \\ k \end{bmatrix}_3$ is the number of k-dimensional hyperplanes in n-attribute SET ($n \leq 7$) containing a given card.

	$k = 0$	$k = 1$	$k = 2$	$k = 3$	$k = 4$	$k = 5$	$k = 6$	$k = 7$
$n = 2$	1	4	1	—	—	—	—	—
$n = 3$	1	13	13	1	—	—	—	—
$n = 4$	1	40	130	40	1	—	—	—
$n = 5$	1	121	1210	1210	121	1	—	—
$n = 6$	1	364	11,011	33,880	11,011	364	1	—
$n = 7$	1	1093	99,463	925,771	925,771	99,463	1093	1

coefficients or, alternatively, the *Gaussian coefficients*, named in honor of the great nineteenth-century mathematician Carl Friedrich Gauss. The q-binomial coefficient $\begin{bmatrix} n \\ k \end{bmatrix}_q$ gives the number of k-dimensional subspaces of an n-dimensional vector space over a finite field with q elements.[11] The general formula looks like this:

$$\begin{bmatrix} n \\ k \end{bmatrix}_q = \frac{(q^n - 1)(q^n - q)(q^n - q^2) \cdots (q^n - q^{k-1})}{(q^k - 1)(q^k - q)(q^k - q^2) \cdots (q^k - q^{k-1})}.$$

To get our formula, just plug in $q = 3$.

We can also restate our hyperplane count $h(n, k)$ in terms of the q-binomials:

$$h(n, k) = 3^{n-k} \begin{bmatrix} n \\ k \end{bmatrix}_3.$$

See table 6.8 for some small values of $\begin{bmatrix} n \\ k \end{bmatrix}_3$. Concentrating on $n = 4$ (the usual game of SET), we note that $\begin{bmatrix} 4 \\ 1 \end{bmatrix}_3 = 40$. This corresponds to the familiar fact that every card is in 40 *SET*s. We also see that $\begin{bmatrix} 4 \\ 2 \end{bmatrix}_3 = 130$ and $\begin{bmatrix} 4 \\ 3 \end{bmatrix}_3 = 40$, i.e., every card is in 130 planes and 40 hyperplanes.

There is a world of interesting mathematics associated with these numbers, and it is beyond the scope of this book to explore that world in

---

[11] If you know some linear algebra, all subspaces contain $\vec{0}$, so counting subspaces corresponds to counting the hyperplanes that contain a given, fixed card. If you don't know linear algebra, don't read this footnote.

detail. But we make a few comments, to possibly motivate you to learn more about it on your own.

1. The numbers in every row of the table are symmetric: they form a palindrome.[12] For example, in the four-attribute game, we see that the number of *SET*s that contain a given card is the same as the number of hyperplanes that contain that card, i.e., $\left[\begin{smallmatrix}4\\1\end{smallmatrix}\right]_3 = \left[\begin{smallmatrix}4\\3\end{smallmatrix}\right]_3 = 40$. In general, you can show the following (exercise 6.6(b)):

$$\begin{bmatrix} n \\ k \end{bmatrix}_q = \begin{bmatrix} n \\ n-k \end{bmatrix}_q.$$

2. For any nonnegative integers q, n, and k (with $k \leq n$), the q-binomial coefficient $\begin{bmatrix} n \\ k \end{bmatrix}_q$ is an integer. In fact, more is true: if we think of q as a variable, then $\begin{bmatrix} n \\ k \end{bmatrix}_q$ is a *polynomial* in q for any legal n and k. For example, with $n = 6$ and $k = 3$, we have

$$\begin{bmatrix} 6 \\ 3 \end{bmatrix}_q = \frac{(q^6 - 1)(q^6 - q)(q^6 - q^2)}{(q^3 - 1)(q^3 - q)(q^3 - q^2)}$$

$$= (q + 1)\left(q^2 + 1\right)\left(q^2 - q + 1\right)\left(q^4 + q^3 + q^2 + q + 1\right)$$

$$= q^9 + q^8 + 2q^7 + 3q^6 + 3q^5 + 3q^4 + 3q^3 + 2q^2 + q + 1.$$

   This always simplifies to a polynomial—for any n, k, and q, you can always cancel all the factors in the denominator with corresponding factors in the numerator.

   Moreover, the coefficients of the polynomial form a symmetric, unimodal sequence: the coefficients of $\begin{bmatrix} 6 \\ 3 \end{bmatrix}_q$ given above are $\{1, 1, 2, 3, 3, 3, 3, 2, 1, 1\}$, a nice palindrome.

3. Plugging in $q = 1$ gives the ordinary binomial coefficients: $\begin{bmatrix} n \\ k \end{bmatrix}_1 = \binom{n}{k}$. (To plug in $q = 1$, you must first reduce the expression for $\begin{bmatrix} n \\ k \end{bmatrix}_q$ to get a polynomial in q.)

We'll use many of the counts from this chapter in chapter 7, where we study probability and expected value.

---

[12] The first name of one of the authors of this book is a palindrome. See if you can figure out which one.

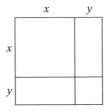

Figure 6.4. A proof without words: $(x + y)^2 = x^2 + 2xy + y^2$.

EXERCISES

EXERCISE 6.1. We found a formula for the number of *SET*s $l(n, k)$ with k attributes the same, that contain a given card, in section 6.4.2. That proof used an incidence count and the formula for the global count $g(n, k)$. Give a direct proof that $l(n, k) = \binom{n}{k}2^{n-k-1}$.

EXERCISE 6.2. In this exercise, you'll show that the global and local percentages of *SET*s with k attributes the same are identical. Fix n and k, where $0 \leq k \leq n-1$, and define $g(n, k)$ and $l(n, k)$ as in section 6.4. Use the formulas for $g(n, k)$ and $l(n, k)$ to show that

$$\frac{g(n, k)}{3^{n-1}(3^n - 1)/2} = \frac{l(n, k)}{(3^n - 1)/2},$$

i.e., the global and local percentages of *SET*s with k attributes the same are equal.

EXERCISE 6.3. In chapter 2, we defined an *interset* to be a collection of four cards that are formed by taking two *SET*s that contain a common card, with that card removed. How many intersets are there in n-attribute SET? [Hint: Adapt the argument given in chapter 2.]

The next two exercises use the *binomial theorem*:

$$(x + y)^n = \sum_{k=0}^{n} \binom{n}{k} x^{n-k} y^k.$$

EXERCISE 6.4. (Binomial theorem, small cases).

a. The $n = 2$ case of the binomial theorem is the familiar $(x + y)^2 = x^2 + 2xy + y^2$. Explain how the picture in figure 6.4 demonstrates this.

b. Try to construct a model that demonstrates the $n = 3$ case:

$$(x + y)^3 = x^3 + 3x^2 y + 3xy^2 + y^3.$$

[Hint: Your model should consist of eight three-dimensional blocks.]

EXERCISE 6.5. Recall that $g(n) = 3^{n-1}(3^n - 1)/2$ is the total number of SETs in the n-attribute version of the game.

a. Show that $g(n + 1) \approx 9 \times g(n)$ for n large. (For example, $g(10) = 9.0003 \times g(9)$.)

b. Use the binomial theorem to show that adding up the number of SETs with k attributes the same gives the total number of SETs:

$$\sum_{k=0}^{n-1} \binom{n}{k} 3^{n-1} 2^{n-k-1} = g(n).$$

c. Verify algebraically that the local versions of part (b) also work: Pick a card C in n-attribute SET. Recall that the number of SETs containing C with k attributes the same is $l(n, k) = \binom{n}{k} 2^{n-k-1}$, and the total number of SETs containing C is $(3^n - 1)/2$. Show that

$$\sum_{k=0}^{n-1} l(n, k) = \frac{3^n - 1}{2}.$$

[Hint: Use part (b).]

EXERCISE 6.6. Prove the following facts about the q-binomial coefficient $\begin{bmatrix} n \\ k \end{bmatrix}_q$. [Non-Hint: There are clever solutions that use facts concerning vector spaces. Attempt to use those at your own risk.]

a. $\begin{bmatrix} n+1 \\ k \end{bmatrix}_q = q^k \begin{bmatrix} n \\ k \end{bmatrix}_q + \begin{bmatrix} n \\ k-1 \end{bmatrix}_q$. [Hint: Use the formula for $\begin{bmatrix} n \\ k \end{bmatrix}_q$ and lots of algebra.]

b. $\begin{bmatrix} n \\ k \end{bmatrix}_q = \begin{bmatrix} n \\ n-k \end{bmatrix}_q$. [Hint: Again, use the formula and cancel like crazy.]

c. Treat q as a variable, and show that $\begin{bmatrix} n \\ k \end{bmatrix}_q$ is a polynomial in q of degree $k(n - k)$. [Hint: Use (a) and math induction on n.]

d. Show that evaluating $\begin{bmatrix} n \\ k \end{bmatrix}_q$ at $q = 1$ gives the binomial coefficient $\binom{n}{k}$. [Hint: Use (a) and math induction.]

EXERCISE 6.7. ("Not" counting) Do the following counts in the n-attribute game:

a. How many collections of three cards are there that do not form a *SET*?
b. How many collections of four cards are there that are not coplanar?
c. Simplify your answers to parts (a) and (b) as much as possible, and then find a nice, general formula for the number of subsets of k cards (where $k \leq n + 1$) that do not contain a *SET*, a plane, a three-dimensional hyperplane, and so on. Such a collection of points is said to be in *general position*. [Hint: See the argument in section 6.5.1.]

EXERCISE 6.8. Pick a card in four-attribute SET, then pick a *SET* containing your card, then pick a plane that contains your *SET*, then pick a hyperplane that contains your plane. This ordered card–*SET*–plane–hyperplane sequence is called a *flag*:

a. How many different flags are there in four-attribute SET?
b. Now find a formula for the number of flags in the n-attribute game.

EXERCISE 6.9. Use incidence counts (or the material developed in section 6.5) to find the following:

a. How many planes contain a given card in the n-attribute game?
b. How many planes contain a given *SET* in the n-attribute game?
c. Why don't we just move on to the big question: How many k-dimensional hyperplanes contain a given d-dimensional hyperplane (where $0 \leq d \leq k \leq n$)?

PROJECTS

PROJECT 6.1. Suppose you split the deck into two piles, with k cards in one pile and $81 - k$ cards in the other. How many *SET*s meet both piles, i.e., how many *SET*s are there that include at least one card in each pile?

There is a quick answer to this problem, but there are some interesting consequences. Let's call a *SET* that contains cards from both sides of the partition a *crossing SET*. To count the number of crossing *SET*s, choose one card from each side. This determines a crossing *SET*, but note that every crossing *SET* is counted twice by this procedure, since the two cards in the *SET* on the same side of the partition will each give rise to the same crossing *SET*.

a. Generalize to the n-attribute game: Partition the 3^n cards in the deck into two parts. Let cr(n, k) be the number of crossing *SET*s with k cards in one part and $3^n - k$ cards in the other. Then

$$\mathrm{cr}(n, k) = \frac{k(3^n - k)}{2}.$$

The idea of looking at *crossing SETs* is due to Macula. He derives a formula equivalent to this in his article "An analysis of the lines in the three-dimensional affine space over F_3," *Ars Combinatoria* **52** (1999), 161–171. (More recently, Jim Vinci posted a similar formula in the Teachers' Corner section of the Set Enterprises website.)

Consequences: You can prove several results from this chapter using this formula.

b. Use the formula for cr(n, k) to show that the number of *SET*s containing a given card is $(3^n - 1)/2$, which agrees with the calculation in this chapter.
c. Use the formula for cr(n, k) to show that the number of *SET*s meeting a given *SET* is $3 \times (3^n - 3)/2$, again agreeing with our previous work.
d. The number of crossing *SET*s does not depend on the configuration of the cards on either side, but only on the number of cards on each side. Use this idea to show the following: Let S be a collection of three cards that do not form a *SET*. Show that the number of *SET*s meeting S is also $3 \times (3^n - 3)/2$.
e. For the usual game with 81 cards, find the maximum and minimum numbers of *SET*s among the 69 cards that remain after the first 12 cards have been dealt. [Hint: Think about the 12 cards that have been dealt.]
f. Choose any two disjoint *SET*s S and T (so $n \geq 2$). Show that the number of other *SET*s that meet either S or T (or both) is $3^{n+1} - 18$. (Note that this formula works whether or not the two *SET*s are coplanar.)
g. Is there a formula for the number of *SET*s that meet three disjoint *SET*s? Here's how you can answer this question.

   i. Suppose we choose the three *SET*s in a plane, as in figure 6.5, and assume we're working in four-attribute SET. (It doesn't matter which three disjoint *SET*s we choose for this problem.)

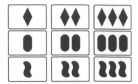

Figure 6.5. Three disjoint *SET*s form a plane.

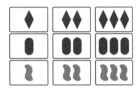

Figure 6.6. Three pairwise coplanar *SET*s.

Show that there are 333 *SET*s that meet one of these three *SET*s (not including the three given *SET*s). Then show that the general formula is $9(3^n - 7)/2$ for the n-attribute game when the three given *SET*s form a plane.

ii. Redo part (i) for the three *SET*s of figure 6.6. Show that this time there are 324 *SET*s that meet one of these three *SET*s (not including the three given *SET*s). This time, the general formula is $9(3^n - 9)/2$. This assumes that every pair of the three *SET*s are coplanar, but they do not form a plane.

Conclude that no general formula exists for the three disjoint *SET*s problem. In the four-attribute game, show that 324 is the smallest number of *SET*s that meet one of three given (disjoint) *SET*s, and 333 is the largest such number.

h. Let S be a collection of $2 \times 3^{n-1}$ cards. Show that S contains at least $3^{n-2}(3^{n-1} - 1)$ *SET*s.

Probability and Statistics

7.1 INTRODUCTION

In chapter 6, we answered all sorts of counting questions involving the n-attribute version of the game. In this chapter, we will use some of those answers to compute probabilities and expected values. Questions about probability and expected value come up naturally in playing SET. We start with a motivating example.

7.1.1 Pick Three Cards. Is it a *SET*?

What is the probability that three randomly chosen cards in the n-attribute game form a *SET*? Recall that we compute the probability of an event by finding the number of ways that event can occur, then dividing by the total number of possibilities.

In chapter 3, we found the answer to this question for the four-attribute game (it's 1/79). To generalize to the n-attribute game, we need two counts from chapter 6. The numerator is the total number of *SET*s in the n-attribute game, $3^{n-1}(3^n - 1)/2$. For the denominator, the total number of ways to choose three cards from the deck is $\binom{3^n}{3} = 3^n(3^n - 1)(3^n - 2)/6$. Then the probability that our three chosen cards form a *SET* is

$$\frac{\text{\# } SET\text{s}}{\text{\# ways to choose 3 cards}} = \frac{3^{n-1}(3^n - 1)/2}{3^n(3^n - 1)(3^n - 2)/6} = \frac{1}{3^n - 2}.$$

It is often the case that there are alternate ways to compute a probability. As in chapter 3, to find the probability that three randomly

Table 7.1.
The probability that three random cards form a *SET* in the n-attribute game for $n \leq 8$.

n	1	2	3	4	5	6	7	8
Probability	100%	14%	4%	1.27%	0.4%	0.14%	0.04%	0.015%

chosen cards form a *SET*, imagine we choose two cards initially. Then only one of the remaining $3^n - 2$ cards will complete a *SET* with these first two cards, so there is a $1/(3^n - 2)$ chance our three cards will form a *SET*.

These probabilities approach 0 rapidly as n increases. For example, in the six-attribute game, the chances of getting a *SET* by picking three random cards is $1/(3^6 - 2) = 0.0013755\ldots \approx 0.14\%$. We list these probabilities for $n \leq 8$ in table 7.1.

This computation tells us that choosing three cards at random in the usual four-attribute game produces a *SET* with probability approximately 1.3%. This means you can't just grab three cards and hope they're a *SET* when you're playing the game—the odds are overwhelmingly against you. On the other hand, this implies that approximately once every 79 times, the three cards you add to a layout should be a *SET*. Since you add three cards around 23 times during a typical game of SET, you might expect this to happen once in every three or four games.

7.1.2 An Expected Value Problem

One reason the game is both challenging and fast paced[1] is that there will usually be a few *SET*s present in any layout of 12 cards. How many *SET*s are there, on average, in the first layout of 12 cards?

In chapter 3, we introduced *linearity of expected value* to answer this question for the four-attribute game. To generalize, if we are playing the n-attribute game and we lay out m cards initially, there are $\binom{m}{3}$ ways to choose three of the cards from the table, and the probability that any one of those three-card subsets forms a *SET* is $1/(3^n - 2)$, from above. That means we can expect to find $\binom{m}{3}/(3^n - 2)$ *SET*s, on average.

---

[1] This depends on who is playing.

TABLE 7.2.
Number of cards needed in initial layout to ensure an average of approximately 2.78 SETs.

# attributes	2	3	4	5	6	7	8	9	10
# cards in initial layout	6	8.5	12	16.9	24	34.2	48.8	70	100.5

Here's one way we can apply this result. Suppose we wanted to play seven-attribute SET, but we still wanted around 2.78 SETs (on average) in the initial layout. How many cards do we need? We must solve the equation

$$\binom{m}{3} \frac{1}{3^7 - 2} = 2.78.$$

Since $\binom{m}{3} = m(m-1)(m-2)/6$, we can use some algebra to rewrite this equation:

$$m^3 - 3m^2 + 2m - 36,445.8 = 0.$$

Solving cubic equations is no picnic,[2] but it's easy enough using a computer algebra system (we used *Mathematica*). Then we find three solutions, which we've rounded to the nearest tenth:

$$m \approx -15.6 - 28.7i, \quad -15.6 + 28.7i, \quad \text{or} \quad 34.2.$$

The first two solutions are *complex numbers* involving the imaginary number $i = \sqrt{-1}$. It would take us too far afield to give any background on what these are and where they arise in mathematics, and we would end up rejecting those two solutions anyway. So we conclude that an initial layout of 34.2 cards would produce an average of around 2.78 SETs.

Of course, it's not possible to lay out fractions of cards.[3] We give the approximate number of cards needed to hit this target for $2 \leq n \leq 10$ in table 7.2.

[2] Assuming you enjoy picnics, but not solving cubic equations. General cubic equations can be solved exactly using the famous *Cardano formulas*, the generalization of the quadratic formula discovered during the early sixteenth century, but not by Cardano.

[3] Well, it's *possible*, but we don't recommend it.

Finally, there's nothing special about the number 2.78. In fact, it might be hard to play the seven-attribute game with 35 cards in an initial layout when there are only 2.78 SETs, on average.[4] See exercise 7.3 for a different approach to this problem.

7.2 STATISTICS FOR THE NUMBER OF *SETS* WITH k ATTRIBUTES THE SAME

What does a typical SET in the n-attribute game look like? We interpret this somewhat vague question in two different ways:

Q1: A SET is picked at random in the n-attribute game. What is the probability that exactly k attributes are the same?

Q2: When a SET is randomly chosen, what is the expected number of attributes that are the same?

We'll start with Q1, finding the probability $P(n, k)$ that our SET has precisely k attributes the same. This is the ratio of the number of all such SETs to the total number of SETs. In chapter 6, we figured out that there are exactly

$$g(n, k) = \binom{n}{k} 3^{n-1} 2^{n-k-1}$$

SETs with k attributes the same. Using this expression for our numerator, our formula for the total number of SETs $3^{n-1}(3^n - 1)/2$ for our denominator, and a little algebra, we can completely answer Q1:

• The probability that a randomly chosen SET will have k attributes the same is

$$P(n, k) = \binom{n}{k} \frac{2^{n-k}}{3^n - 1}.$$

---

[4] Honestly, if you're playing the seven-attribute game, there will be bigger challenges than this one.

TABLE 7.3.
The probabilities a random *SET* will have *k* attributes the same.

	k = 0	*k* = 1	*k* = 2	*k* = 3	*k* = 4	*k* = 5	*k* = 6	*k* = 7
n = 1	100%	—	—	—	—	—	—	—
n = 2	50%	50%	—	—	—	—	—	—
n = 3	30.7%	46.2%	23.1%	—	—	—	—	—
n = 4	20%	40%	30%	10%	—	—	—	—
n = 5	13.2%	33.1%	33.1%	16.5%	4.1%	—	—	—
n = 6	8.8%	26.4%	33%	22%	8.2%	1.6%	—	—
n = 7	5.8%	20.5%	30.7%	25.6%	12.8%	3.8%	0.6%	—
n = 8	3.9%	15.6%	27.3%	27.3%	17%	6.8%	1.7%	0.2%

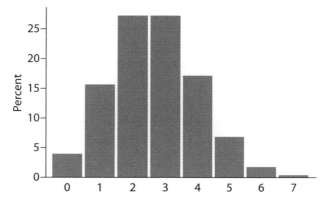

Figure 7.1. The percentages of *SET*s with *k* attributes the same in the eight-attribute game. Note that $k = 2$ and $k = 3$ are tied for the maximum value. The average is around 2.66—this is the "balance point" on the horizontal axis for the graph.

We give some data for $P(n, k)$ in table 7.3. Note that the "nice" percentages of 20%, 40%, 30%, and 10% that appear in row 4 of the table agree with our earlier computations. So, for instance, in the usual four-attribute game, there is a 40% chance your *SET* will have one attribute the same and three different.

If you like to visualize your data (and you should), then check out the graph for the eight-attribute game in figure 7.1.

For Q2, we need to figure out the expected value for the number of attributes that are the same. Let's start with four-attribute SET, the game we all know and love. There are 216 *SET*s with zero attributes the same,

432 with one attribute the same, 324 with two the same, and 108 with three the same. Then we get

$$\text{expected value} = \frac{216 \times 0 + 432 \times 1 + 324 \times 2 + 108 \times 3}{1080} = \frac{13}{10} = 1.3.$$

What we're after is an exact formula for the expected number of attributes that are the same for the general, n-attribute game. We will need the binomial theorem (introduced in exercises 6.4 and 6.5):

Binomial Theorem

$$(x + y)^n = \binom{n}{0}x^n + \binom{n}{1}x^{n-1}y + \binom{n}{2}x^{n-2}y^2 + \cdots + \binom{n}{n}y^n.$$

Now we're ready to find a formula. Here's a guide:

1. Call the expected value a_n. Then the expected value is given by

$$\frac{0 \times g(n, 0) + 1 \times g(n, 1) + 2 \times g(n, 2) + \cdots + (n-1) \times g(n, n-1)}{3^{n-1}(3^n - 1)/2}.$$

2. We can use the formula $g(n, k) = \binom{n}{k}3^{n-1}2^{n-k-1}$ from chapter 6 and some algebra to simplify this sum:

$$a_n = \frac{0 \times \binom{n}{0}2^n + 1 \times \binom{n}{1}2^{n-1} + 2 \times \binom{n}{2}2^{n-2} + \cdots + (n-1) \times \binom{n}{n-1}2^1}{3^n - 1}.$$

At this point, we will use sigma notation to rewrite the sum on the top of the fraction more compactly:

$$0 \times \binom{n}{0}2^n + 1 \times \binom{n}{1}2^{n-1} + \cdots + (n-1) \times \binom{n}{n-1}2^1$$
$$= \sum_{k=0}^{n-1} k\binom{n}{k}2^{n-k}.$$

If the notation $\sum_{k=0}^{n-1} k\binom{n}{k}2^{n-k}$ is new to you, it might look scary.[5] But the big $\sum$ just means "add up a bunch of terms," where each term corresponds to a value of k from 0 to $n-1$.

3. To simplify this formula, we'll use the binomial theorem.[6] We concentrate on the top of the fraction, $T = \sum_{k=0}^{n-1} k\binom{n}{k}2^{n-k}$, and find an expression (using the binomial theorem and calculus[7]) that (almost) gives us T.

   i. First, plug $y = 2$ into the binomial theorem $(x + y)^n = \sum_{k=0}^{n} \binom{n}{k}x^k y^{n-k}$ to get

$$(x + 2)^n = \sum_{k=0}^{n} \binom{n}{k} x^k 2^{n-k}.$$

   ii. Now, take the derivative of each side with respect to x:

$$n(x + 2)^{n-1} = \sum_{k=0}^{n} k\binom{n}{k} x^{k-1} 2^{n-k}.$$

   iii. Now plug in $x = 1$. This gives us

$$n(1 + 2)^{n-1} = \sum_{k=0}^{n} k\binom{n}{k} 1^{k-1} 2^{n-k}.$$

     Simplifying, we get

$$n \times 3^{n-1} = \sum_{k=0}^{n} k\binom{n}{k} 2^{n-k}.$$

   iv. We're getting close. The expression $\sum_{k=0}^{n} k\binom{n}{k}2^{n-k}$ is *almost T*. The only problem is that we've got the $k = n$ term, which does not appear in T (because we can't have a *SET* with n attributes

---

[5] Might?

[6] Or a computer algebra package. But it's more fun to do it by hand, don't you think?

[7] If you haven't taken calculus, just trust us. Or mosey on over to exercise 7.2 for an alternate approach.

Table 7.4.
The expected value for the number of attributes that are the same in a randomly
chosen *SET* in the n-attribute game.

n	1	2	3	4	5	6	7	8	9	10
Expected value	0	0.5	0.92	1.3	1.65	2.00	2.33	2.67	3.00	3.33

the same—this would be the same card three times):

$$n \times 3^{n-1} = \left(\sum_{k=0}^{n-1} k \binom{n}{k} 2^{n-k} \right) + n \binom{n}{n} 2^{n-n} = T + n.$$

Solving for T gives us $T = n(3^{n-1} - 1)$.

v. Now that we have an expression for T, we can use it to get a formula for the expected value a_n. The expected number of attributes that are the same is

$$a_n = \frac{T}{3^n - 1} = \frac{n(3^{n-1} - 1)}{3^n - 1}.$$

An alternate (and very slick) derivation that makes use of the linearity of expected value appears in exercise 7.1. Table 7.4 summarizes the expected number of attributes that are the same for a randomly chosen *SET*.

The expected number of attributes that are the same rises slowly. In fact, adding an attribute increases the expected number of attributes that are the same by roughly $\frac{1}{3}$ when n increases by 1. We'll look more closely at this in the next section.

7.3 COIN FLIPPING, SET, AND THE CENTRAL LIMIT THEOREM

The two questions Q1 and Q2 from section 7.2 give us some information about what a typical *SET* looks like. But we can reformulate these questions. Doing so will eventually lead us to an unexpected connection between the n-attribute game and coin flipping.

Our reformulation of Q1 and Q2 involves picking cards from the deck. Suppose you and your friend Sumiko each pick a card from the n-attribute deck, and then you compare your cards.

Q3: What is the probability that your two cards will be the same in exactly k attributes?

Q4: What is the expected number of attributes that will be the same for your two cards?

But these are the same questions we considered in section 7.2: Q1 and Q3 have the same answer for any n and any value of $k < n$, and Q2 and Q4 are also identical for any n. Why? There are two reasons, both of which are familiar:

- Every pair of cards determines a unique *SET* (the fundamental theorem).
- If two cards in a *SET* have exactly k attributes the same, then all the cards in the *SET* have k attributes the same.

Think of this as the legacy of the fundamental theorem: when we know two cards, we know everything we need to know about the third card in the *SET* they determine.

The advantage of considering questions Q3 and Q4 is that we are choosing just two cards from a deck, *without replacement*. People who study probability distinguish between drawing cards *with replacement* or *without replacement*. It's always easier to do the computations in the former case: once you replace the card, you can assume the second draw is *independent* from the first, and that means lots of nice things happen. But we're working *without replacement* here: once you've drawn your card, you keep it. Sumiko's choice *depends* on what you've picked: she can't pick your card.

7.3.1 Approximate Solutions

What happens if we solve the problem *with replacement*? We'll get the wrong answer, but it will be very close to the exact answers we found in section 7.2. In fact, when n is reasonably large, the difference between the exact answer and this approximation will be very, very small. You'll see.

To do the problem *with replacement*, assume you pick a card at random from the deck, then you replace it, and then Sumiko picks her card.

APPROXIMATE SOLUTION TO Q3

1. There are k choices (among the n attributes) for the attributes that will be the same for your two cards. Choose these k attributes in $\binom{n}{k}$ ways.
2. Now choose the other attributes. For each of the $n - k$ attributes that are different, there are two choices. For instance, if your card is red, and color was not one of the attributes that are the same, then Sumiko's card must be green or purple. Since there are two choices for each of these $n - k$ attributes, we have 2^{n-k} cards that Sumiko can choose that are different in these $n - k$ attributes.
3. Finally, since there were 3^n cards Sumiko could have selected, the probability her card will match yours in precisely k attributes is $\binom{n}{k}\frac{2^{n-k}}{3^n}$.

How good is our approximation? From section 7.2, we have the exact answer of

$$P(n, k) = \binom{n}{k} \frac{2^{n-k}}{3^n - 1}.$$

Our approximation just replaces the $3^n - 1$ term in the denominator of the formula for $P(n, k)$ by 3^n. The difference between the exact and approximate answers is trivial when n is large. For example, when $n = 8$, the approximations for all possible values of k agree with the corresponding exact values in at least four decimals. See table 7.5.

What happens in the four-attribute game? In this case, our approximations are still quite good. See table 7.6 for the data.

You might notice something from the numbers in tables 7.5 and 7.6: all of the approximations are slightly smaller than the exact values. Here's why: When we choose cards with replacement, the chances that Sumiko's card matches your card in all n attributes is $(\frac{1}{3})^n$. But Sumiko's card *must be different* from your card, so this probability should

TABLE 7.5.
The exact and approximate probabilities a random *SET* will have *k* attributes the same in eight-attribute SET.

k	Approximate	Exact	Difference
0	0.0390184	0.0390244	5.9479×10^{-6}
1	0.156074	0.156098	0.0000237917
2	0.273129	0.273171	0.0000416355
3	0.273129	0.273171	0.0000416355
4	0.170706	0.170732	0.0000260222
5	0.0682823	0.0682927	0.0000104089
6	0.0170706	0.0170732	2.60222×10^{-6}
7	0.00243865	0.00243902	3.717458×10^{-7}

TABLE 7.6.
Approximate and exact probabilities that two cards agree in 0, 1, 2, or 3 attributes in four-attribute SET.

# attributes the same	0	1	2	3
Approximate	19.8%	39.5%	29.6%	9.9%
Exact	20%	40%	30%	10%

equal 0. This extra case is the difference between working with replacement and without replacement, and it means that the percentages in the approximation add up to (slightly) less than 100%. But this error is very small when *n* is large.

This approach works for the expected values, too.

APPROXIMATE SOLUTION TO Q4

Q4: What is the expected number of attributes that will be the same for your two cards?

Here's the solution, again assuming you *replace* your card before Sumiko chooses her card.

1. There is a $\frac{1}{3}$ chance Sumiko's card will match yours in the first attribute, a $\frac{1}{3}$ chance it will match in the second attribute, and so on.

TABLE 7.7.

Expected value (exact and approximate) for the number of attributes that are the same in a randomly chosen *SET* in the *n*-attribute game.

n	Approx.	Exact	Difference	n	Approx.	Exact	Difference
1	0.333333	0	0.333333	6	2	1.99451	0.00549451
2	0.666667	0.5	0.166667	7	2.333333	2.3312	0.0021348
3	1	0.923077	0.0769231	8	2.666667	2.66585	0.000813008
4	1.333333	1.3	0.0333333	9	3	2.9997	0.000304847
5	1.666667	1.65289	0.0137741	10	3.333333	3.33322	0.000112902

2. Then the expected number of attributes that are the same is simply

$$\underbrace{\frac{1}{3} + \frac{1}{3} + \cdots + \frac{1}{3}}_{n \text{ times}} = \frac{n}{3}.$$

Back in section 7.2, we found the exact answer $a_n = n(3^{n-1}-1)/(3^n-1)$. Again, the approximation of $n/3$ is excellent. To see this, let's rewrite a_n:

$$a_n = \frac{n(3^{n-1} - 1)}{3^n - 1} = \frac{n}{3} \times \left(\frac{3^n - 3}{3^n - 1} \right).$$

When n is large, the term $(3^n - 3)/(3^n - 1)$ is very close to 1, so a_n is very close to $n/3$. We give the exact and approximate values for the expected values in table 7.7 for $n \leq 10$.

7.3.2 Coin Flipping

Before we connect our approximate solutions to Q3 and Q4 to coin flipping, we take a brief time-out to review some background material we will need. Here's a problem you might encounter in a probability course:

Question: Flip a coin 10 times. What is the probability you got exactly 6 heads?

A standard solution might look like this: First, there are $\binom{10}{6} = 210$ ways to choose the six times the coin came up heads. We could (in theory) list all 210 possibilities:

HHHHHHTTTT, HHHHHTHTTT, . . . , TTTTHHHHHH.

Each of these 210 possibilities occurs with probability $(\frac{1}{2})^{10}$. Since these events are *disjoint*, the answer is $210 \times \left(\frac{1}{2^{10}}\right) \approx 20.5\%$.

In order to relate coin flipping to Q3 and Q4, we need to use a weighted coin.[8]

COIN-FLIPPING FORMULATION FOR Q3

Question: Flip a weighted coin n times, where heads comes up with probability $\frac{1}{3}$ and tails comes up with probability $\frac{2}{3}$. What is the probability you got exactly k heads?

The solution should look familiar: Choose the k places for heads in $\binom{n}{k}$ ways. Then the probability that a sequence of flips came up heads k times and tails $n - k$ times is $(\frac{1}{3})^k(\frac{2}{3})^{n-k}$, and this probability is the same for any sequence of k heads and $n - k$ tails. Putting the pieces together gives

$$\binom{n}{k}\left(\frac{1}{3}\right)^k\left(\frac{2}{3}\right)^{n-k} = \binom{n}{k}\frac{2^{n-k}}{3^n}.$$

This agrees with our answer to Q3 from section 7.3.1.

COIN-FLIPPING FORMULATION FOR Q4

For Q4, the expected number of heads in n flips will be $n/3$. Again, this matches our approximate solution to Q4 from section 7.3.1. The principal advantage to using the coin-flipping model is that the percentages of flips that give k heads (for $0 \le k \le n$) follow the (very well-known) *binomial distribution* from statistics.

---

[8] If someone offers you the chance to play a betting game with a weighted coin, don't.

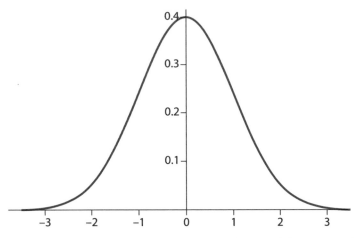

Figure 7.2. The standard normal curve has mean 0 and standard deviation 1. The total area under the curve is 1.

7.3.3 The Normal Curve

Lots of data in statistics follow the familiar bell curve distribution, called the *normal* curve[9] by statisticians. For example, the distribution of the heights of all adults in the United States follows this curve quite closely. Normal curves are all similar to each other: we can always obtain one from another by rescaling.

Normal curves are completely characterized by two numbers: the mean and the standard deviation. The standard deviation tells you how spread out the curve is. See figure 7.2 for a graph of the normal curve with mean 0 and standard deviation 1.

We mention one more fact about this distribution. Most of the data are clustered rather closely to the mean. In fact, about 68% of the data are within one standard deviation of the mean and 95% are within two standard deviations. We'll use this in the next section, when we connect coin flipping to the normal curve.

---

[9] The word *normal* is overused in mathematics, with definitions in geometry (*normal vectors*), analysis (*normal numbers*), algebra (*normal subgroups* and *normal field extensions*), topology (*normal spaces*), and the list goes on. The name is apt in statistics—many, many different distributions follow this curve, approximately, so it's "normal."

7.3.4 The Bell Curve and Coin Flipping

When you flip a (weighted or unweighted) coin n times, the probability distribution for the number of heads (or tails) is very well behaved. When n is reasonably large, this distribution approaches a normal curve.

Why is this true? It follows from one of the most important results in probability and statistics, the *central limit theorem*. This theorem essentially tells us that whenever you flip a coin (weighted or not) repeatedly and count the frequencies for the number of heads, the bar graph for the number of heads approaches a bell-shaped curve. The more flips, the better the approximation. (In fact, the central limit theorem tells us much more, but it's beyond the scope of this book to explore this topic.)

As a general rule, if n is the number of flips and p is the probability that your weighted coin comes up heads, you can use the normal approximation whenever both $np > 10$ and $n(1-p) > 10$. In our case, we know $p = \frac{1}{3}$, so we would be confident using this approximation for $n > 30$.

We summarize the main point of this section with an example.

Question: In 60-attribute SET, what percentage of all *SET*s have between 15 and 25 attributes the same? A quick glance at the graph in figure 7.3 should convince you the answer is "most of them."

We'll do this problem twice: once exactly, and once using our coin-flipping model.

1. **Exact:** First, the probability that our randomly chosen *SET* has exactly k attributes the same in n-attribute SET is $P(n, k) = \binom{n}{k} 2^{n-k}/(3^n - 1)$. Then the exact answer to our problem is a sum:

$$\binom{60}{15} \frac{2^{45}}{(3^{60} - 1)} + \binom{60}{16} \frac{2^{44}}{(3^{60} - 1)} + \cdots$$

$$+ \binom{60}{25} \frac{2^{35}}{(3^{60} - 1)} = 86.9026\ldots\%.$$

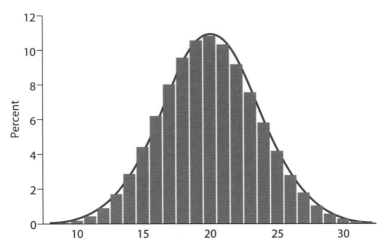

Figure 7.3. The percentages of *SET*s with *k* attributes the same in the 60-attribute game. The red curve is a bell-shaped normal curve, with mean 20 and standard deviation (approximately) 3.65.

2. **Coin-flipping approximation:** Next, we approximate our answer by using the binomial, coin-flipping model. This time, our sum is

$$\binom{60}{15}\frac{2^{45}}{3^{60}} + \binom{60}{16}\frac{2^{44}}{3^{60}} + \cdots + \binom{60}{25}\frac{2^{35}}{3^{60}} = 86.9026\ldots\%.$$

How close are these two answers? We find the approximation and the exact answers agree in the first 28 decimals; this difference is smaller than any human could possibly care about. (The normal curve also gives us a way to estimate the answer—see exercise 7.6.)

We conclude this section with a discussion of how the normal approximation can give us information about what most *SET*s look like when *n* is large. Using a standard fact about normal curves, we know that approximately 95% of all *SET*s will have the number of attributes the same within two standard deviations of the mean, and virtually all of the *SET*s (around 99.7%) will have this number within three standard deviations.

You can see this for yourself in figure 7.4, which shows the total number of *SET*s with *k* or fewer attributes the same in a *cumulative distribution function*. The graph of such a function always moves

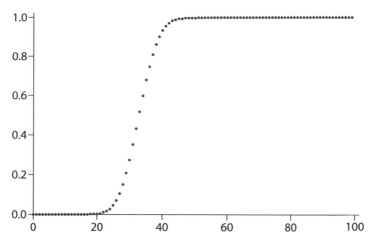

Figure 7.4. The *cumulative distribution function* for the percentage of *SET*s with *k* or fewer attributes the same in the 100-attribute game. Virtually all of the *SET*s have between 20 and 40 attributes the same.

from 0 (representing none) to 1 (representing all) on the vertical axis as you move from left to right on the horizontal axis.

This means that if you're playing 100-attribute SET,[10] don't waste too much time looking for *SET*s with lots of attributes the same, or lots of attributes different. The probability that a randomly chosen *SET* will have between 19 and 47 attributes the same is greater than 99%. (Actually, don't waste your time at all on the 100-attribute game. From our earlier calculations, you'd need to have around 1.45689×10^{16} cards on the table for there to be three *SET*s, on average.)

7.4 MEDIAN AND MODE: A PREVIEW

We now know the expected value for the number of attributes that are the same in a randomly chosen *SET* in the *n*-attribute version of the game—it's very close to *n*/3. From the data analysis point of view, this tells us that the *mean* or *average* number of attributes shared is approximately *n*/3. What is the *median* value, i.e., the number *m* of attributes where half the *SET*s have no more than *m* attributes the same

[10] Where do you keep your deck of 3^{100} cards?

TABLE 7.8.
The mean, median, and mode for the number of attributes that are the same in a *SET*.

# attributes	1	2	3	4	5	6	7	8	9	10
Mean	0	0.5	0.9	1.3	1.7	2.0	2.3	2.7	3.0	3.3
Median	0	0.5	1	1	2	2	2	3	3	3
Mode	0	0,1	1	1	1,2	2	2	2,3	3	3

and half the *SET*s have no less than m attributes the same? What is the *mode*, i.e., the number of attributes that is the most likely to occur? These three numbers, the *mean*, *median*, and *mode*, are standard ways to measure the "middle" of the data.

For 10-attribute SET, the mean is about 3.3, the median is 3, and the mode is also 3 (there are more *SET*s with 3 attributes the same than any other number). One consequence of the close approximation of the data by the normal curve is that, like the mean, both the median and mode are very close to $n/3$.

If you like this sort of thing, you might enjoy working through the details of project 7.1, where you can find formulas for the median and mode. We give the mean, median, and mode for the number of attributes that are the same in n-attribute SET for $n \leq 10$ in table 7.8.

We encourage you to ask questions, to search for patterns, and to convince yourself those patterns always hold. For instance, for what values of n are the median and mode equal? When are there two equal modes? Is $n = 2$ the only time the median isn't an integer? (The median is not an integer when $n = 2$ because exactly half the *SET*s have no attributes the same and half have one attribute the same.)

Finally, in four-attribute SET, we've known for some time that 20% of the *SET*s have no attributes the same, 40% have one attribute the same, 30% have two attributes the same, and 10% have three attributes the same. Recall that $P(n, k)$ is the probability that a randomly chosen *SET* will have k attributes the same. Putting these probabilities in decreasing order, we have

$$P(4, 1) > P(4, 2) > P(4, 0) > P(4, 3).$$

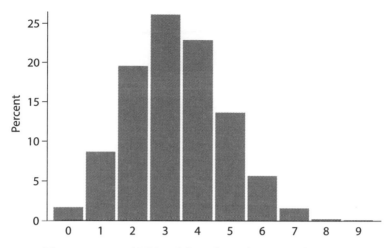

Figure 7.5. The percentages of *SET*s with k attributes the same in the 10-attribute game. Order the percentages from highest to lowest.

What happens for the general n? For instance, glancing at the graph for $n = 10$ in figure 7.5 should convince you that those probabilities are ordered as follows (where, for convenience, we write p_k instead of $P(10, k)$):

$$p_3 > p_4 > p_2 > p_5 > p_1 > p_6 > p_0 > p_7 > p_8 > p_9.$$

Does this pattern[11] always hold? We encourage you to investigate!

EXERCISES

EXERCISE 7.1. In section 7.2, we found the formula for a_n, the expected number of attributes that are the same in a randomly chosen *SET* in the n-attribute game. In this exercise, we outline a quick derivation that takes advantage of the linearity of expected value. For two cards A and B, let $X_i = 1$ if A and B are the same in attribute i, and 0 otherwise.

a. Show that $P(X_i = 1) = (3^{n-1} - 1)/(3^n - 1)$ by counting the number of cards B that agree with card A in attribute i.

b. Explain why $E(X_i) = P(X_i = 1)$, where $E(X_i)$ is the expected value of X_i.

---

[11] A perfectly reasonable question would be, "What pattern?"

c. Now use the linearity of expected value to show

$$a_n = E(X_1) + E(X_2) + \cdots + E(X_n) = \frac{n(3^{n-1} - 1)}{3^n - 1}.$$

EXERCISE 7.2. In our derivation of a formula for the average number of attributes a random *SET* has the same (section 7.2), we used calculus to show

$$\sum_{k=0}^{n} k \binom{n}{k} 2^{n-k} = n3^{n-1}.$$

In this exercise, we'll show that this is true by counting the same thing in two different ways. First, we need a story.

A big SET tournament is taking place, and n people have entered. There will be one champion, who receives a nice savings bond, and everyone else will receive a certificate. There are three types of certificates: A level, B level, and C level.

---

a. With this backstory, let's count the number of possible outcomes of the tournament. Show that there are $n3^{n-1}$ possible outcomes by first picking the champion, then giving out the certificates to everyone else.
b. Now let's count the number of possible outcomes slowly. First, we'll choose the champion and the A level people: choose k people, pick one as the champion, and give A level certificates to the remaining $k - 1$ of them. Next, distribute B and C level certificates to the remaining $n - k$ people. Show that, counting this way, the number of possible tournament outcomes is $\sum_{k=1}^{n} k \binom{n}{k} 2^{n-k}$. (Note that the $k = 0$ case is not counted here, but that case contributes nothing to the sum.)
c. Conclude that the identity is true. This is a standard example of *combinatorial reasoning*, where we count the same thing in two different ways.

EXERCISE 7.3. Maybe one reason the four-attribute game is so good is that 2.78 *SET*s (on average) in the first 12 cards is the "right" ratio of *SET*s to cards, namely $2.78/12 \approx 0.23$.

a. How many cards would we need to lay out in seven-attribute SET to ensure this ratio is the same? [Hint: Let m be the number of cards needed in the initial layout to make the expected number of *SET*s equal to $0.23m$. Then solve the equation $\binom{m}{3}\frac{1}{3^7-2} = 0.23m$. Note that this reduces to a quadratic equation.]

b. Now solve part (a) for the general n-attribute game. Make this as general as possible: Let a be the target ratio for the expected number of *SET*s in the initial layout, and determine the initial number of cards needed in the n-attribute game to achieve this target. Your answer should depend on both a and n. [Hint: As in part (a), there will be a quadratic equation in m to solve. Check your answer by setting $a = 0.23$ and $n = 4$. You should get $m \approx 12$.]

EXERCISE 7.4. Suppose we play four-attribute SET, but we remove two-dimensional planes, instead of lines (lines are the usual *SET*s, of course). How many cards would you need to lay out to have a reasonable number of planes present, on average? (Define "reasonable" to be 3.)

EXERCISE 7.5. Let $p(n, k) = \frac{1}{3^n-1} \sum_{i=0}^{k} \binom{n}{i} 2^{n-i}$ be the percentage of *SET*s with k or fewer attributes the same.

a. Let $q(n, k) = (3^n - 1)p(n, k)$. Show that $q(n, k)$ satisfies the following recursion:

$$q(n + 1, k) = 2q(n, k) + q(n, k - 1).$$

b. Use part (a) to show

$$p(n + 1, k) = \frac{3^n - 1}{3^{n+1} - 1}(2p(n, k) + p(n, k - 1)).$$

Thus, when n is reasonably large (say $n \geq 10$), we have

$$p(n + 1, k) \approx \tfrac{2}{3}p(n, k) + \tfrac{1}{3}p(n, k - 1).$$

(Note that this recursion is exact for the coin-flipping approximation.)

c. Use part (b) to show $p(n, k) < p(n + 1, k + 1)$ for $0 \leq k \leq n - 1$ and n "sufficiently large."

d. Use part (b) to show $p(n, k) > p(n + 1, k)$. As above, we require n to be large enough for the approximation in part (b) to be quite good. (For instance, there is a greater percentage of *SET*s with seven or fewer attributes the same when $n = 11$ than there is when $n = 12$.)

EXERCISE 7.6. We wish to use the normal curve to approximate the percentage of *SET*s with between 15 and 25 attributes the same in the 60-attribute game. We know that the answer is approximately 86.9% from our work in section 7.3.4.

a. Show that the mean is 20 and the standard deviation is (approximately) 3.65. [Hint: The standard deviation for the binomial distribution is $\sqrt{p(1-p)n}$, where p is the probability of heads and n is the number of flips.]

b. Use a table or a computer to look up the area under the curve between $X_1 = 15$ and $X_2 = 25$.

c. The approximation in (b) is not very accurate; the error is around 4%. One reason for this discrepancy is that vertical lines at $x = 15$ and $x = 25$ cut the vertical bars of the histogram in half, so our approximation is too small. We can fix this problem by using $X_1 = 14.5$ and $X_2 = 25.5$ for our vertical lines on the normal curve (this is usually called the *continuity correction*). Repeat part (a) using these new X-values. (Your new approximation should be accurate to within a tenth of a percent.)

EXERCISE 7.7. We know that the mean for the number of attributes the same for a randomly chosen *SET* in the n-attribute game is approximately $n/3$. The standard deviation for this distribution is approximately $\sqrt{2n}/3$. Show that the exact value for the standard deviation is

$$\sqrt{\frac{3^{2n-3}(2n)\left(3^n - 2n - 1\right)}{\left(3^n - 1\right)\left(3^{2n-1} - 3^{n-1} - 2\right)}}.$$

[Hint: Look up the definition of standard deviation somewhere, then use what you know about the mean. To simplify your formula, either spend lots of effort doing some algebra, or use a computer algebra system.]

PROJECTS

PROJECT 7.1. (Median and mode) In this project, you'll work through the various properties of the sequence for the number of *SET*s with k attributes the same. See table 7.9.

TABLE 7.9.
The number of *SET*s with k attributes the same.

	$k = 0$	$k = 1$	$k = 2$	$k = 3$	$k = 4$
$n = 1$	1	—	—	—	—
$n = 2$	6	6	—	—	—
$n = 3$	36	54	27	—	—
$n = 4$	216	432	324	108	—
$n = 5$	1296	3240	3240	1620	405

TABLE 7.10.
Maximum number of *SET*s for n-attribute SET ($n \leq 15$) occurs when k attributes are the same for the given k.

# attributes	1	2	3	4	5	6	7	8	9	10	11	12
Mode	0	0,1	1	1	1,2	2	2	2,3	3	3	3,4	4

We will need our formula for the number of *SET*s with k attributes the same:

$$g(n, k) = \binom{n}{k} 3^{n-1} 2^{n-k-1}.$$

We know that the graph of our data closely matches the normal curve (see figure 7.3), so the median and mode should both be very close to the mean, which is approximately $n/3$. The problem is that the median and mode will both be integers, so this gives us the median and the mode only when n is divisible by 3. For the other cases, we need some more work.

Part 1: Mode

For ordinary, four-attribute SET, 40% of all the *SET*s have exactly one attribute the same and three different, and this percentage is larger than the alternatives. This means that the mode occurs for $k = 1$. All the modes for $n \leq 12$ are given in table 7.10.

Let's figure out the mode for n-attribute SET. Here's the deal. Fix n, and recall (from chapter 6) that

$$g(n, k) = \binom{n}{k} 3^{n-1} 2^{n-k-1}$$

is the number of SETs with exactly k attributes the same. The goal of this part of the project is to show that the mode occurs when $k = \lfloor \frac{n}{3} \rfloor$ (this is the *floor* function—round down if n is not divisible by 3).

a. Let $m = \lfloor \frac{n}{3} \rfloor$. Show

   i. $g(n, 0) < g(n, 1) < \cdots < g(n, m)$, and
   ii. $g(n, m) \geq g(n, m+1) > g(n, m+2) > \cdots > g(n, n-1)$.

   [Hint: Consider the ratio $g(n, k+1)/g(n, k)$. Note that we allow the possibility that $g(n, m) = g(n, m+1)$—see part (b).] (A sequence that increases to a maximum, then decreases, is called *unimodal*.)

b. When is there a tie for the maximum? Use part (a) to show that the maximum occurs at two adjacent k values precisely when $n + 1$ is a multiple of 3, i.e., when $n = 2, 5, 8, 11, \ldots$.

c. What about the minimum value? Show that the minimum occurs when $k = n - 1$, i.e., when all but one of the attributes are the same. [Hint: Use part (a) to compare $g(n, 0)$ and $g(n, n-1)$.]

d. What percentage of all the SETs are represented by the maximum? Let's look at some data, in table 7.11.

   It appears that the percentage of the SETs that occur at the mode represents a smaller and smaller percentage of all the SETs, as n increases. Our goal is to show that this is true—the percentage of SETs at the mode tends to 0 as n increases. This will have the flavor of an estimate from real analysis.

   i. Call the fraction of SETs at the mode p_n. Show that

   $$p_n = \binom{n}{m} \frac{2^{n-m}}{3^n - 1},$$

   where $m = \lfloor \frac{n}{3} \rfloor$. [Hint: Use the expressions we already have for the mode, the number of SETs $g(n, m)$ with m attributes the same, and the total number of SETs in n-attribute SET.]

   ii. To estimate p_n, we will need to estimate $\binom{n}{m}$. This, in turn, will use *Stirling's formula*:

   $$n! \approx \frac{1}{\sqrt{2\pi}} \left(\frac{n}{e}\right)^n.$$

TABLE 7.11.
The percentage of the *SET*s that occur at the mode.

n	20	40	60	80	100
Mode	6	13	20	26	33
Percentage	18.2%	13.3%	10.1%	9.4%	8.4%

TABLE 7.12.
How good is the approximation?

n	100	200	300	400	500
Mode	33	66	100	133	166
Exact p_n	0.0843827	0.0596057	0.0488128	0.0422834	0.0377872
Approx. $3/\sqrt{4\pi n}$	0.0846284	0.0598413	0.0488603	0.0423142	0.037847

To make the calculations a little easier, we'll use $k = \frac{n}{3}$ instead of $\lfloor \frac{n}{3} \rfloor$ for our estimates. (This won't make a difference for large values of n.) Show that

$$p_n \approx \frac{3}{\sqrt{4\pi n}}.$$

[Hint: Write $\binom{n}{k} = \frac{n!}{k!(n-k)!}$, and then use Stirling's formula to estimate $n!$, $k!$, and $(n-k)!$, where $k = n/3$. Then do a ton of algebra.]

iii. Conclude that $p_n \to 0$ as $n \to \infty$.

How good is this approximation? Rather than go through a detailed analysis, we just look at some data, in table 7.12.

Part 2: Median

Half the data are greater than or equal to the median and half are smaller or equal. Here, we need to find the smallest value of k so that the number of *SET*s with k or fewer attributes the same is at least half the total number of *SET*s, i.e.,

$$g(n, 0) + g(n, 1) + \cdots + g(n, k) \geq \frac{3^{n-1}(3^n - 1)}{4}.$$

Unfortunately, there are no nice, easy-to-digest formulas for $g(n, 0) + g(n, 1) + \cdots + g(n, k)$. Exact formulas for this sum involve *hypergeometric functions*, which are beyond the scope of the book. But it turns out we can still find a nice formula for the first time this sum exceeds 50%.

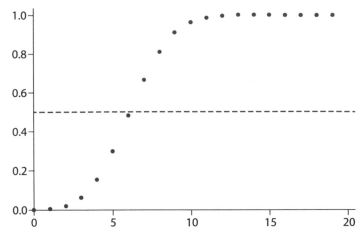

Figure 7.6. The proportion of *SET*s with k or fewer attributes the same in 20-attribute SET, where $k = 0, 1, \ldots, 19$. The dashed line represents 50% of the total number of *SET*s. Since this line is above the data point for $k = 6$ but below the $k = 7$ point, the median is 7.

The graph in figure 7.6 shows how to find the median for the number of attributes that are the same in 20-attribute SET. We give the percentage of *SET*s with fewer than k attributes the same, for all possible values of k.

When $n = 20$, we find that 47.9% of all *SET*s have six or fewer attributes the same, and 66.1% of all *SET*s have seven or fewer attributes the same. This tells us that the median is $k = 7$.

a. Compute the median for $n = 1, 2, 3, 4$. Check that your answers match those given in table 7.8.
b. Let med(n) be the median for the number of attributes that are the same in n-attribute SET. Show med(n) $\leq$ med($n + 1$). [Hint: In exercise 7.5, note that if $p(n, k) > 0.5$, then $k \geq$ med(n). Use part (c) of that exercise.]
c. Show med($n + 1$) $\leq$ med(n) $+ 1$, i.e., the median increases by at most 1 in moving from the n-attribute game to the $(n + 1)$-attribute game. [Hint: Exercise 7.5(d).]

Vectors and Linear Algebra

8.1 INTRODUCTION

Way back in chapter 1, we introduced coordinates for the SET cards, so that every card could be written as a 4-tuple of numbers modulo 3. We continue to use the scheme in table 8.1.

Thinking about the cards as vectors has been useful throughout this book. In fact, the property that three cards form a *SET* if and only if the sums of the individual coordinates are all 0 (mod 3) gave us our first meaingful connection between the game and coordinates. We explore the connections between vectors and the game in more detail here, concluding with affine transformations, which provide a way to measure the symmetry of the deck. In particular, we describe precisely those transformations that also preserve the number of different attributes of all *SET*s.

8.2 PARALLEL *SET*S

We begin by showing how we can use coordinates to define parallel *SET*s. While most of our examples are in the usual four-dimensional game, these ideas generalize easily to n dimensions.

8.2.1 Parallel *SET*s

Parallel *SET*s were introduced in chapter 5. Recall that two *SET*s are parallel if they are coplanar and they don't intersect. We'll use vector translation to define parallel *SET*s here, and then we'll show

TABLE 8.1.
Assignment of coordinates to cards.

Attribute	Value		Coordinate
Number	3, 1, 2	↔	0, 1, 2
Color	green, purple, red	↔	0, 1, 2
Shading	empty, striped, solid	↔	0, 1, 2
Shape	diamonds, ovals, squiggles	↔	0, 1, 2

Figure 8.1. Two parallel *SET*s. We created the *SET* on the right by adding the vector $\vec{w} = (1, 0, 1, 2)$ to the coordinates of each card in the *SET* on the left.

that this definition agrees with the two descriptions of parallel *SET*s we introduced in chapter 5. (If vectors are new to you, we hope the examples will provide you with some geometric and algebraic intuition.)

We demonstrate our procedure for creating a *SET* parallel to a given *SET* via an example. First, choose a *SET*, as on the left in figure 8.1.

To do any algebraic operations, we need to find the coordinates for each of the cards in this *SET*:

$$1 \text{ Green Empty Diamond} \quad \mapsto \quad (1,0,0,0),$$
$$1 \text{ Purple Solid Oval} \quad \mapsto \quad (1,1,2,1),$$
$$1 \text{ Red Striped Squiggle} \quad \mapsto \quad (1,2,1,2).$$

Next, choose any vector $\vec{w}$. For the example, we select $\vec{w} = (1, 0, 1, 2)$. (This vector is completely arbitrary here. It also corresponds to a card in the deck, but we'll ignore that.)

Now add $\vec{w}$ to each of the three vectors corresponding to the cards in the *SET*:

$$(1, 0, 0, 0) + (1, 0, 1, 2) = (2, 0, 1, 2),$$
$$(1, 1, 2, 1) + (1, 0, 1, 2) = (2, 1, 0, 0),$$
$$(1, 2, 1, 2) + (1, 0, 1, 2) = (2, 2, 2, 1).$$

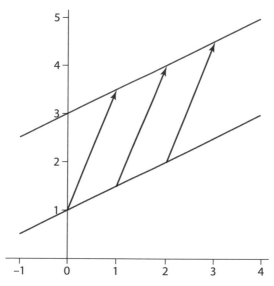

Figure 8.2. Two parallel lines in the Euclidean plane. You can create the top line by adding the same vector to every point on the bottom line.

Finally, figure out what cards correspond to these three new vectors. Those cards are shown on the right in figure 8.1.

This operation may be familiar, if you've studied vector addition in the Euclidean plane. Adding the same vector to every point on a line produces a line with the same slope, so the new line is parallel to the original. See figure 8.2.

Furthermore, even though we used our coordinates repeatedly in this procedure, the fact that the two *SET*s are parallel *does not depend* on how we assign the coordinates. For example, we could change our assignment of numbers to colors to red ↦ 0, green ↦ 1, and purple ↦ 2. Then, as long as we're consistent with this new assignment, everything we've done here is still valid.

This might seem a little surprising, but it's important: we want parallelism to be a property of the *SET*s, not the specific coordinates. Said another way,

If you and your friend Theano each devised your own, different, coordinate assignments, the procedure above would give the same pairs of parallel *SET*s.

Figure 8.3. We want to find the *SET* containing the card on the right parallel to the *SET* on the left.

8.2.2 Vectors and the Parallel Postulate

Adding a vector to a *SET* gives us a parallel *SET*. Let's connect this idea to the parallel postulate:

> Given a *SET* and a card not in that *SET*, there is a unique *SET* containing that card and parallel to the given *SET*.

How can we accomplish this using vector addition? Consider the *SET* on the left in figure 8.3 and the card shown on the right.

The coordinates for 3 Purple Striped Diamonds are $\vec{v} = (0, 1, 1, 0)$, so what vector $\vec{w}$ should we add to the coordinates of the cards in the *SET*? We have a choice: choose any of the cards in the *SET*, and subtract its coordinates from $(0, 1, 1, 0)$. (It turns out that we'll end up with the same parallel *SET*, regardless of which card we choose from the *SET*. See exercise 8.1.)

So, for example, take the first vector, $(1, 0, 0, 0)$. Then

$$\vec{w} = (0, 1, 1, 0) - (1, 0, 0, 0) = (-1, 1, 1, 0) = (2, 1, 1, 0) \quad (\text{mod } 3).$$

Now, add $\vec{w} = (2, 1, 1, 0)$ to each of the three vectors that correspond to the cards in our *SET*:

$$(1, 0, 0, 0) + (2, 1, 1, 0) = (0, 1, 1, 0),$$

$$(1, 1, 2, 1) + (2, 1, 1, 0) = (0, 2, 0, 1),$$

$$(1, 2, 1, 2) + (2, 1, 1, 0) = (0, 0, 2, 2).$$

What two new cards are produced by this procedure? We get 3 Red Empty Ovals (from $(0, 2, 0, 1)$) and 3 Green Solid Squiggles (from $(0, 0, 2, 2)$), as in figure 8.4. (This procedure must include the given card $\vec{v}$ in the parallel *SET*, because $\vec{v}_1 + (\vec{v} - \vec{v}_1) = \vec{v}$. In this case, we have $\vec{v} = (0, 1, 1, 0)$.)

Figure 8.4. We found the parallel *SET*.

8.2.3 Vectors and Cyclic Attributes

In section 5.3, we gave two different ways to define two parallel *SET*s. We now explain why the vector description above agrees with the cycle description of parallel *SET*s from chapter 5. Recall, two *SET*s are parallel if you can place the two *SET*s so that, for all the attributes that are not the same, the cyclic ordering is the same. Also recall that the cyclic orderings (a, b, c), (b, c, a), and (c, a, b) are considered the same.

First, let's check that the two *SET*s in figure 8.1 have all the attributes cycling in the same way:

- Number: Each card in the first *SET* has 1 symbol, while each card in the second *SET* has 2 symbols.
- Color: In both *SET*s, from left to right, the color cycle is (green, purple, red).
- Shading: In both *SET*s, from left to right, the shading cycle is (empty, solid, striped).
- Shape: Again, in both *SET*s, from left to right, the shape cycle is (diamonds, ovals, squiggles).

Why does vector addition preserve these cycles? Here's a brief explanation. Call the three vectors in the original *SET* $\vec{v}_1$, $\vec{v}_2$, and $\vec{v}_3$, and write

$$\vec{v}_1 = (a_1, b_1, c_1, d_1), \qquad \vec{v}_2 = (a_2, b_2, c_2, d_2), \qquad \vec{v}_3 = (a_3, b_3, c_3, d_3).$$

Now create a parallel *SET* by adding a vector $\vec{w} = (r, s, t, u)$ to each of the three vectors in the *SET*.

- If the three expressions of an attribute are the same in the first *SET*, they're the same in the second. In the example, we have $a_1 = a_2 = a_3 = 1$, so, setting $r = 1$, we have $a_1 + r = a_2 + r = a_3 + r = 2$.

- If the three expressions are different in the first *SET*, then they remain different and cycle the same way in the second. This time, looking at the third coordinate, we have the numbers $c_1 = 0$, $c_2 = 2$, and $c_3 = 1$. Adding $t = 1$ to each, we have $c_1 + t = 1$, $c_2 + t = 0$, and $c_3 + t = 2$.

  This process will always produce an equivalent cycle. Since it doesn't matter which number in a cycle comes first, adding the same number to each number in a cycle (working mod 3) will simply shift the cycle, producing the same cyclic ordering of the attributes.

8.2.4 Direction Vectors and Parallel *SET*s

Given two *SET*s, how can we use vectors to determine if they're parallel? Our solution involves *direction vectors*. Here's how it works.

1. First, compute the direction vector $\vec{d}$ of a *SET* by taking the difference between any two vectors corresponding to cards in the *SET*. For the *SET*s in figure 8.1, we use the first two cards for each *SET* to get direction vectors:

$$SET\ 1: \vec{d_1} = (1, 1, 2, 1) - (1, 0, 0, 0) = (0, 1, 2, 1),$$

$$SET\ 2: \vec{d_2} = (2, 1, 0, 0) - (2, 0, 1, 2) = (0, 1, 2, 1).$$

2. Then the two *SET*s are parallel if and only if $\vec{d_1} = \vec{d_2}$ or $\vec{d_1} = 2\vec{d_2}$. In this case, we have $\vec{d_1} = \vec{d_2}$.

Direction vectors are determined only up to (scalar) multiples. For instance, had we used $\vec{v_1}$ and $\vec{v_3}$ to determine our direction vector for our first *SET*, we would have

$$\vec{d_1} = \vec{v_3} - \vec{v_1} = (1, 2, 1, 2) - (1, 0, 0, 0) = (0, 2, 1, 2) = 2(0, 1, 2, 1) = 2\vec{d_2}.$$

But the direction vector does not depend on the assignment of coordinates: if Theano assigned vectors to the cards in a different way, she would get the same direction vectors we got. (See exercise 8.2.)

Figure 8.5. We can add three different vectors to the *SET* on the left to produce the parallel *SET* on the right.

***Takeaway Message*:** Two *SET*s are parallel if and only if their corresponding direction vectors are multiples of each other.

8.2.5 The Number of Parallel *SET*s

Chapters 6 and 7 were devoted to counting things. In that spirit, we ask the following question:

- How many *SET*s are parallel to a given *SET*?

Suppose you have a *SET* in the usual four-dimensional game. To find a parallel *SET*, we could add any of the $3^4 = 81$ possible vectors $\vec{w}$ to coordinates corresponding to our three cards. But this will overcount the number of parallel *SET*s. To see why, let's look at an example.

Choose the *SET* on the left in figure 8.5, with cards corresponding to the vectors $\vec{v}_1 = (2, 0, 2, 0)$, $\vec{v}_2 = (1, 1, 0, 1)$, and $\vec{v}_3 = (0, 2, 1, 2)$. Let $\vec{d} = \vec{v}_2 - \vec{v}_1 = (2, 1, 1, 1)$ be a direction vector for this *SET*. Then adding $\vec{d}$ to each of the vectors in this *SET* will simply reorder the cards in our *SET*:

$$\vec{v}_1 + \vec{d} = (2, 0, 2, 0) + (2, 1, 1, 1) = (1, 1, 0, 1) = \vec{v}_2,$$

$$\vec{v}_2 + \vec{d} = (1, 1, 0, 1) + (2, 1, 1, 1) = (0, 2, 1, 2) = \vec{v}_3,$$

$$\vec{v}_3 + \vec{d} = (0, 2, 1, 2) + (2, 1, 1, 1) = (2, 0, 2, 0) = \vec{v}_1.$$

In fact, there are three vectors that leave the original *SET* unchanged when we add them to the three vectors $\vec{v}_1$, $\vec{v}_2$, and $\vec{v}_3$, namely $\vec{d} = (2, 1, 1, 1)$, $2\vec{d} = (1, 2, 2, 2)$, and $0 \times \vec{d} = (0, 0, 0, 0)$. So, given any *SET*, there are three vectors we could add to the vectors representing the cards in that *SET* that "do nothing."

How many different vectors could we add to the vectors belonging to the original *SET* to produce another, parallel *SET*? Symbolically, if the parallel *SET* has representing vectors $\vec{u}_1$, $\vec{u}_2$, and $\vec{u}_3$, how many different vectors $\vec{w}$ are there that satisfy

$$\{\vec{v}_1 + \vec{w},\ \vec{v}_2 + \vec{w},\ \vec{v}_3 + \vec{w}\} = \{\vec{u}_1, \vec{u}_2, \vec{u}_3\}$$

in some order?

Returning to our example, the three cards in the *SET* on the right in figure 8.5 have coordinates $\vec{u}_1 = (0, 1, 1, 0)$, $\vec{u}_2 = (2, 2, 2, 1)$, and $\vec{u}_3 = (1, 0, 0, 2)$. Now let $\vec{w}_1 = \vec{u}_1 - \vec{v}_1 = (1, 1, 2, 0)$. Adding $\vec{w}_1$ to the vectors $\vec{v}_1$, $\vec{v}_2$, and $\vec{v}_3$ gives

$$\vec{v}_1 + \vec{w}_1 = (2, 0, 2, 0) + (1, 1, 2, 0) = (0, 1, 1, 0) = \vec{u}_1,$$

$$\vec{v}_2 + \vec{w}_1 = (1, 1, 0, 1) + (1, 1, 2, 0) = (2, 2, 2, 1) = \vec{u}_2,$$

$$\vec{v}_3 + \vec{w}_1 = (0, 2, 1, 2) + (1, 1, 2, 0) = (1, 0, 0, 2) = \vec{u}_3.$$

But we could also have used $\vec{w}_2 = \vec{w}_1 + \vec{d}$ or $\vec{w}_3 = \vec{w}_1 + 2\vec{d}$, where $\vec{d} = (1, 2, 2, 2)$ is the direction vector above. Adding either $\vec{w}_2$ or $\vec{w}_3$ to each of $\vec{v}_1$, $\vec{v}_2$, and $\vec{v}_3$ in turn would produce the same three vectors $\vec{u}_1$, $\vec{u}_2$, and $\vec{u}_3$, in some order. This gives three separate ways to transform the first *SET* into the second:

$+\vec{w}_1$				$+\vec{w}_2$				$+\vec{w}_3$	
$\vec{v}_1$	$\mapsto$	$\vec{u}_1$		$\vec{v}_1$	$\mapsto$	$\vec{u}_2$		$\vec{v}_1$	$\mapsto$	$\vec{u}_3$
$\vec{v}_2$	$\mapsto$	$\vec{u}_2$		$\vec{v}_2$	$\mapsto$	$\vec{u}_3$		$\vec{v}_2$	$\mapsto$	$\vec{u}_1$
$\vec{v}_3$	$\mapsto$	$\vec{u}_3$		$\vec{v}_3$	$\mapsto$	$\vec{u}_1$		$\vec{v}_3$	$\mapsto$	$\vec{u}_2$

We can now use this information to answer the original question. Given a *SET*, we can produce a parallel *SET* by adding three different vectors. Thus, there are $81/3 = 27$ *SET*s parallel to a given *SET*. But this includes the original *SET*, which is parallel to itself. So this leaves 26 *other SET*s parallel to the given *SET*.

Instead of using vectors, we could have answered this question using the Parallel Postulate. Since every card is in exactly one *SET* parallel to the given *SET*, the entire deck must be the disjoint union of the *SET* and its parallel *SET*s. This means that in the n-dimensional version of the game, there must be 3^{n-1} *SET*s parallel to a given *SET* (including the given *SET*).

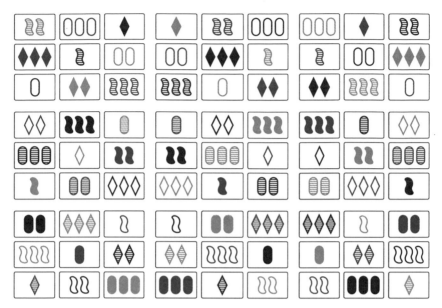

Figure 8.6. The entire deck, again. For a challenge, find all 26 *SET*s parallel to the *SET* consisting of 1 Green Solid Diamond, 2 Green Striped Squiggles, and 3 Green Empty Ovals.

Adding the same vector to every vector in a *SET* (or the entire deck) is an example of an *affine transformation*. We'll study these in detail in section 8.4.

We end this section with a familiar picture. You can see the entire deck in figure 8.6. Pick a *SET*, then try to find all the *SET*s parallel to your *SET*. For example, if you choose the three cards in the top row of the upper left plane (2 Green Striped Squiggles, 3 Red Empty Ovals, and 1 Purple Solid Diamond), then the *SET*s parallel to this *SET* are the 27 horizontal *SET*s in the picture (including the *SET*, which is parallel to itself). Note that each of these *SET*s has all four attributes different. Exercise 8.4 asks you to find more *SET*s parallel to a given *SET* in this figure.

8.2.6 Using Parallel *SET*s to Create a Plane

Parallelism is intimately connected to planarity. More precisely, we can make the following statement.

- Two non-intersecting *SET*s are parallel if and only if there is a plane containing both *SET*s.

TABLE 8.2.
Every plane can be partitioned into three parallel *SET*s.

$\vec{v}_1$	$\vec{v}_2$	$\vec{v}_3$
$\vec{v}_1 + \vec{w}$	$\vec{v}_2 + \vec{w}$	$\vec{v}_3 + \vec{w}$
$\vec{v}_1 + 2\vec{w}$	$\vec{v}_2 + 2\vec{w}$	$\vec{v}_3 + 2\vec{w}$

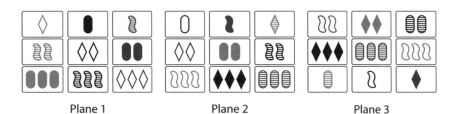

Plane 1　　　　　　Plane 2　　　　　　Plane 3

Figure 8.7. Three planes. Plane 1 is parallel to either plane 2 or plane 3 (but not the other). Which one?

Here's why two parallel *SET*s live in a unique plane. If you have two parallel *SET*s, write them using coordinates:

$$SET \ 1{:}\{\vec{v}_1, \vec{v}_2, \vec{v}_3\}, \qquad SET \ 2{:} \{\vec{v}_1 + \vec{w}, \vec{v}_2 + \vec{w}, \vec{v}_3 + \vec{w}\}.$$

Then it's easy to complete the plane—the last three cards will be $\vec{v}_1 + 2\vec{w}$, $\vec{v}_2 + 2\vec{w}$, and $\vec{v}_3 + 2\vec{w}$. Then these nine cards, as in table 8.2, form a plane, because the lines in AG(2, 3) exactly match the collections of three vectors that sum to $\vec{0}$ (mod 3). (There are several sums to check, but all of this works out nicely.)

Conversely, if we are given a plane, how do we know that the vectors can be arranged to look like table 8.2? This is the point of exercise 8.5, where you are asked to connect the way we constructed a plane in chapter 5 (from three cards that do not form a *SET*) to the construction in table 8.2.

8.2.7 Parallel Planes, Hyperplanes, Etc.

When are two planes parallel? Here's a quiz: There are three planes in figure 8.7. Plane 1 is parallel to either plane 2 or plane 3, but not both. See if you can figure out which pair are parallel.

TABLE 8.3.
The vector representations of the cards in plane 1.

(1,0,0,0)	(1,1,2,1)	(1,2,1,2)
(2,0,1,2)	(2,1,0,0)	(2,2,2,1)
(0,0,2,1)	(0,1,1,2)	(0,2,0,0)

TABLE 8.4.
Adding $\vec{w} = (1, 2, 0, 2)$ to every vector in table 8.3.

(2,2,0,2)	(2,0,2,0)	(2,1,1,1)
(0,2,1,1)	(0,0,0,2)	(0,1,2,0)
(1,2,2,0)	(1,0,1,1)	(1,1,0,2)

We will determine when two planes are parallel algebraically, exactly as we did for *SET*s. Given a plane, we can construct a parallel plane as follows: As before, first convert the nine cards in our plane to vectors. Next, choose a vector $\vec{w}$ to add to each of your nine vectors. Finally, convert those nine vector sums back to cards in the deck.

For example, choose plane 1 in figure 8.7. This plane is represented by the nine vectors[1] in table 8.3. Then adding $\vec{w} = (1, 2, 0, 2)$ to each vector in plane 1 gives the vectors in table 8.4.

You can now check that the nine vectors in table 8.4 correspond to the cards in plane 3 of figure 8.7 (although they aren't in the same order). We have a winner! (You can also check that there is no vector that will transform plane 1 to plane 2 in this way.)

We can also use this procedure to define parallel *hyperplanes* of any dimension. So the entire (four-dimensional) deck can be partitioned into 27 parallel *SET*s, or 9 parallel planes, or 3 parallel hyperplanes. For example, the 27 cards in the first 9 × 3 column of figure 8.6 form a hyperplane, and the other two hyperplanes parallel to it are the other two 9 × 3 columns in the figure.

---

[1] A useful app for your electronic device, or special glasses, would "see" the cards and translate them into their vector representations. (This is an excellent place for a joke using the words "set cards," "vectors," and "The Matrix." We leave this joke to the interested reader.)

Figure 8.8. Five cards at the end: find the missing card.

Figure 8.9. The End Game. What you expected (left) and what you got (right).

8.3 ERROR CORRECTING, VECTORS, AND SET*

We introduced the End Game in chapter 1, and we revisited it in chapter 4. Suppose you play a game and come up with the collection of five cards in figure 8.8.

Being a SET expert at this point, you quickly announce, "3 Red Solid Squiggles!" (shown on the left in figure 8.9). Dramatically, you flip the missing card, and you see the card on the right in the figure.

You didn't get the right card! You were expecting a card with three symbols, but the card you turned over had only one symbol. How could that have happened? Someone must have made a mistake during the game—they took three cards that were not a *SET*. Let's use vectors to analyze what went wrong.

The End Game works because vectors for the cards at the end of the game must sum to $\vec{0}$. This follows from the argument we gave in chapter 4: the whole deck sums to $\vec{0}$, and each of the *SET*s taken during the play of the game sums to $\vec{0}$. If the End Game fails, there is only one possible explanation: there must be at least one "*SET*" that got taken that was not a *SET*. We can think of the End Game as a parity check: it can detect if a mistake was made during the play of the game. (Parity checks are common: UPC symbols and bank routing transit numbers both use check digits to detect errors.)

Now what? Having found that a mistake was made, you look through all of the *SET*s taken, and you discover the non-*SET* in figure 8.10.

---

* Note: Some of the information in this section first appeared in the chapter "Error detection and correction using SET" (by two of the authors of this book), in *The Mathematics of Various Entertaining Subjects: Research in Recreational Math*, J. Beineke and J. Rosenhouse eds., Princeton University Press, 2015.

Figure 8.10. This non-*SET* was taken during the game.

Figure 8.11. Let's hope this non-*SET* was never taken during a game.

Figure 8.12. Two canceling non-*SET*s.

Notice that the mistaken *SET* was incorrect in number. This is visible in the End Game: the difference between the card we expected (3 Red Solid Squiggles) and the card we got (1 Red Solid Squiggle) was also in number. This is not a coincidence. Using vectors, the coordinate that caused the non-*SET* in figure 8.10 to fail to be a *SET* will affect the same coordinate in the End Game.

What if your predicted card differs from the card you turn over in more than one attribute? There are two possibilities:

1. One non-*SET* was taken, but it failed to be a *SET* in more than one attribute, as in figure 8.11. This kind of mistake is rare among experienced players.
2. More than one non-*SET* was taken during the game.

Can the End Game detect *all* mistakes made during the play of the game? No! It is possible that two mistakes are made that cancel each other out, as in figure 8.12. In this case, the End Game works perfectly because the two non-*SET*s together sum to $\vec{0}$. (In fact, these could be six cards left at the end of the game, for the same reason.)

8.3.1 Hamming Weight

Coding theory is the study of error-correcting codes, which are used for transmitting data in a variety of applications. For instance, codes are essential when images are sent to earth from space. They are also used in reading mp3 files and transmitting data over a wireless network, and in electronic devices of all sorts.

For SET, here's how coding theory enters the picture. Recall that when we had the non-*SET* in figure 8.10, the mistake was in number. What happens if you add the coordinates of the cards in that figure? The three cards have coordinates $(0, 1, 2, 1)$, $(0, 2, 0, 1)$, and $(2, 0, 1, 1)$ in the order they are shown in the figure. The (mod 3) sum of these coordinates is $(2, 0, 0, 0)$, and the one nonzero entry is in position 1, which corresponds to number.

If S is *any* collection of three cards, define the *Hamming weight* of S to be the number of nonzero coordinates (it doesn't matter if the nonzero element is a 1 or a 2) in the (mod 3) sum of the coordinates of the cards of S.

Look at figure 8.11 for an example of three cards that are not a *SET*. The coordinates for the cards in the figure are $(0, 2, 2, 0)$, $(2, 1, 2, 1)$, and $(1, 1, 1, 2)$. Their (mod 3) sum is $(0, 1, 2, 0)$, which has Hamming weight 2, because there are two nonzero coordinates. And, indeed, this non-*SET* is wrong in two attributes. Further, the nonzero entries of the sum are in positions 2 and 3, which correspond to color and shading, exactly the attributes that were wrong in the non-*SET*. Treating each coordinate separately, the argument is the same as the argument we used in justifying the End Game.

8.3.2 Correcting the Error

Suppose we play the End Game and detect an error. Can we then correct it?[2] That will depend on what we mean by "correct."

Look again at our example of a non-*SET* in figure 8.11, where the sum of the coordinates is $(0, 1, 2, 0)$. Then

$$(0, 2, 2, 0) + (2, 1, 2, 1) + (1, 1, 1, 2) + 2 \times (0, 1, 2, 0) =$$

$$(0, 0, 0, 0) \quad (\text{mod } 3).$$

We call the vector $2 \times (0, 1, 2, 0) = (0, 2, 1, 0)$ the *error vector*, E.

---

[2] Coding theory is all about correcting errors, because you want your picture of Pluto to look perfect.

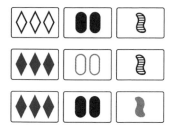

Figure 8.13. Mistake fixed, three times. Each row is potentially the "corrected" version of the non-*SET* in figure 8.11. Note that each of these *SET*s has two cards in common with the non-*SET*.

Here is a strategy for fixing the mistake. First, find the coordinates for the cards: $A = (0, 2, 2, 0)$, $B = (2, 1, 2, 1)$, and $C = (1, 1, 1, 2)$. Since $A + B + C + E = (0, 0, 0, 0) \pmod 3$, we can add E to each of A, B, and C, in turn.

This will fix the error, producing three different *SET*s:

$$\{A + E, B, C\}, \qquad \{A, B + E, C\}, \qquad \{A, B, C + E\}.$$

Now $A + E = (0, 1, 0, 0) = 3$ Purple Empty Diamonds, $B + E = (2, 0, 0, 1) = 2$ Green Empty Ovals, and $C + E = (1, 0, 2, 2) = 1$ Green Solid Squiggle. This gives three genuine *SET*s—see figure 8.13.

One consequence: We can't tell which card was "wrong" in figure 8.11. In reality, no single card is "wrong," but you may get that impression if you find a non-*SET* during the game. This is a psychological issue, not a mathematical one.

8.4 AFFINE EQUIVALENCE: ALL *SET*S ARE THE SAME

In this chapter, we've relied on the vector representations of cards and *SET*s, but we haven't yet needed more sophisticated linear algebra. To analyze the symmetry of the deck, we will need to understand special functions, called *affine transformations*.

Our goal in this section is to explain how all *SET*s are geometrically the "same." This is also valid for planes and hyperplanes, but we need matrices to understand why.

For our purposes, a *matrix* is a square array of n^2 numbers,[3] arranged in n rows and n columns. We will want to think of an $n \times n$ matrix as a *function*, where the input and output are vectors with n entries. We'll use $n = 4$ for the applications to SET, but this generalizes to any positive integer n.

It's almost always easiest to understand a new idea via an example. Let M be the following 4×4 matrix:

$$M = \begin{pmatrix} 1 & 2 & 0 & 1 \\ 1 & 0 & 2 & 0 \\ 0 & 0 & 1 & 2 \\ 2 & 0 & 1 & 2 \end{pmatrix}.$$

Let $\vec{v}$ be the following vector:

$$\vec{v} = \begin{pmatrix} 1 \\ 0 \\ 1 \\ 2 \end{pmatrix}.$$

(We will want to write our vectors as columns throughout this section, but they still correspond to cards in the SET deck.)

We compute the matrix product $M\vec{v}$, where all of our arithmetic will be done mod 3.

Then we get

$$M\vec{v} = \begin{pmatrix} 1 & 2 & 0 & 1 \\ 1 & 0 & 2 & 0 \\ 0 & 0 & 1 & 2 \\ 2 & 0 & 1 & 2 \end{pmatrix} \begin{pmatrix} 1 \\ 0 \\ 1 \\ 2 \end{pmatrix} = \begin{pmatrix} 0 \\ 0 \\ 2 \\ 1 \end{pmatrix}.$$

---

[3] For other people's purposes, matrices aren't necessarily squares. But we don't need to be those people here.

Think of the vector $\vec{v} = (1, 0, 1, 2)$ as the input here, and the matrix M transforms $\vec{v}$ into the output vector $(0, 0, 2, 1)$ (where we write both $\vec{v}$ and the output vector $M\vec{v}$ horizontally for convenience).

The definition of a linear transformation using a matrix appears below.[4] Remember, the matrix is just an array of numbers, and the vectors are columns of numbers. The n^2 entries of M are indexed $a_{11}, a_{12}, \ldots, a_{nn}$. Then $M\vec{v}$ is computed as follows:

$$
\begin{pmatrix}
a_{11} & a_{12} & \cdots & a_{1n} \\
a_{21} & a_{22} & \cdots & a_{2n} \\
\vdots & \vdots & & \vdots \\
a_{n1} & a_{n2} & \cdots & a_{nn}
\end{pmatrix}
\begin{pmatrix}
v_1 \\
v_2 \\
\vdots \\
v_n
\end{pmatrix}
=
\begin{pmatrix}
a_{11}v_1 + a_{12}v_2 + \cdots + a_{1n}v_n \\
a_{21}v_1 + a_{22}v_2 + \cdots + a_{2n}v_n \\
\vdots \\
a_{n1}v_1 + a_{n2}v_2 + \cdots + a_{nn}v_n
\end{pmatrix}.
$$

If you haven't seen matrix multiplication before, this definition may look strange. Although the output vector on the right-hand side of the equals sign looks forbidding, each entry is just the sum of a bunch of products of numbers that appear on the left-hand side. To learn more about linear algebra (and maybe get more out of this chapter), we have two suggestions: Sheldon Axler wrote a book with the promising title *Linear Algebra Done Right*, and if you like free books, Robert Beezer has an online text called *A First Course in Linear Algebra*, available at http://linear.ups.edu/html/fcla.html.

As we mentioned above, we think of the vector $\vec{v}$ as the input and the new vector $M\vec{v}$ as the output of an operation. Applying this to SET, we can view the matrix M as a recipe for permuting the cards. In the example, our input vector $(1, 0, 1, 2)$ corresponds to the card 1 Green Striped Squiggle, and the output vector $(0, 0, 2, 1)$ is the card 3 Red Solid Ovals. Again, we've written these vectors horizontally, purely for convenience.[5]

Matrix multiplication is a *linear transformation*. Linear transformations have the following very important property: if a and b are constants and $\vec{v}_1$ and $\vec{v}_2$ are vectors, then $M(a \cdot \vec{v}_1 + b \cdot \vec{v}_2) = a \cdot M(\vec{v}_1) + b \cdot M(\vec{v}_2)$. (We note that most functions do not have this property. A

---

[4] If this is scary, take a deep breath. It will all be over soon.
[5] Our convenience.

common mistake many students make when they learn algebra is to implicitly assume all functions are linear.[6])

Why do we care about transformations of vectors? We are interested in the permutations of the deck that preserve *SET*s, planes, and hyperplanes. The transformations we will develop in section 8.4.1 are precisely the functions that have these properties. We'll also figure out the total number of symmetries of the entire deck; it will be a number we've met before.

8.4.1 Affine Transformations

Matrix multiplication is important in defining our transformations of the deck, but it is also unnecessarily restrictive. Here's the problem: If M is any 4×4 matrix, then we always have the matrix product $M\vec{0} = \vec{0}$. That means that the card corresponding to $\vec{0}$ (3 Green Empty Diamonds) must be fixed by every linear transformation. But this card is arbitrary, of course, and we want all cards to be treated the same by our transformations.

To fix this problem, we will define an *affine transformation*. We'll stick to the $n = 4$ case here (corresponding to the usual game of SET), but all of what we are about to do works for a general n.

Let M be a 4×4 matrix and let $\vec{b}$ be a 4-tuple. Define a function on the *SET* of all vectors of length 4 by

$$T(\vec{v}) = M\vec{v} + \vec{b}.$$

Let's try this for the matrix M above and the vector $\vec{b} = (0, 1, 1, 2)$. For practice, we apply the transformation to three vectors that form a *SET*: $\vec{v}_1 = (1, 1, 0, 2)$, $\vec{v}_2 = (0, 1, 2, 1)$, and $\vec{v}_3 = (2, 1, 1, 0)$. For $\vec{v}_1 = (1, 1, 0, 2)$, we find $M\vec{v}_1 + \vec{b} = (2, 2, 2, 2)$. We think of this as mapping the card corresponding to $\vec{v}_1$ (1 Purple Empty Squiggle) to the card corresponding to $T(\vec{v}_1) = (2, 2, 2, 2)$ (2 Red Solid Squiggles). You can see what T does to the other two vectors in table 8.5.

Note that the three cards $\vec{v}_1$, $\vec{v}_2$, and $\vec{v}_3$ form a *SET*, and so do $T(\vec{v}_1)$, $T(\vec{v}_2)$, and $T(\vec{v}_3)$ (see figure 8.14). (You might also note that these *SET*s

---

[6] We wish we had a dollar for every time we saw the mistake $\sqrt{x + y} = \sqrt{x} + \sqrt{y}$. We would then have at least $10.

TABLE 8.5.
Applying the transformation $T(\vec{v}) = M\vec{v} + \vec{b}$ to three cards that form a *SET*.

Input	Card	Output	Card
(1,1,0,2)	1 Purple Empty Squiggle	(2,2,2,2)	2 Red Solid Squiggles
(0,1,2,1)	3 Purple Solid Ovals	(0,2,2,0)	3 Red Solid Diamonds
(2,1,1,0)	2 Purple Striped Diamonds	(1,2,2,1)	1 Red Solid Oval

Figure 8.14. T maps the first *SET* to the second.

have different numbers of attributes the same: the first *SET* has one attribute the same, and the second has two.) This always works for affine transformations:

- Affine transformations map *SET*s to *SET*s.

We can prove this. Assume that $\vec{v}_1$, $\vec{v}_2$, and $\vec{v}_3$ form a *SET*. Then we know $\vec{v}_1 + \vec{v}_2 + \vec{v}_3 = \vec{0}$. We must show that the same thing is true for $T(\vec{v}_1)$, $T(\vec{v}_2)$, and $T(\vec{v}_3)$, i.e.,

$$T(\vec{v}_1) + T(\vec{v}_2) + T(\vec{v}_3) = \vec{0}.$$

Now

$$T(\vec{v}_1) + T(\vec{v}_2) + T(\vec{v}_3) = (M(\vec{v}_1) + \vec{b}) + (M(\vec{v}_2) + \vec{b}) + (M(\vec{v}_3) + \vec{b})$$

$$= M(\vec{v}_1 + \vec{v}_2 + \vec{v}_3) + 3\vec{b}$$

$$= M\vec{0} + \vec{0}$$

$$= \vec{0}.$$

Affine transformations also preserve intersets, planes, and hyper-planes (see exercise 8.8). We conclude that affine transformations preserve all the things we want them to preserve. (In chapter 9, we'll see they preserve even some things you might not have thought about, like collections of cards with no *SET*s.)

Figure 8.15. Find an affine transformation taking the first *SET* to the second.

For our transformations to be permutations of the deck, we will want the matrix to be *nonsingular*. This will mean that the transformation is *one-to-one*, so we never have two input cards mapped to the same output card. There are a variety of ways to see if a matrix is nonsingular, but we won't need them now.

Here's a provocative statement:

- All *SET*s are the same, up to the symmetry of affine transformations.

Here's what we mean: if you take any two *SET*s, there is some affine transformation T that takes one to the other. (In fact, there are quite a few, as we'll see.) We say that any two *SET*s are *affinely equivalent*.

It's time for an example. Consider the two *SET*s with vector representations $\{(1, 0, 0, 0), (2, 1, 2, 1), (0, 2, 1, 2)\}$ and $\{(1, 1, 2, 0), (1, 1, 2, 1), (1, 1, 2, 2)\}$. These *SET*s are pictured in figure 8.15. We're going to find an affine transformation that takes the first *SET* to the second. Write the first *SET* as $\{\vec{v}_1, \vec{v}_2, \vec{v}_3\}$ and the second as $\{\vec{w}_1, \vec{w}_2, \vec{w}_3\}$.

Let

$$M = \begin{pmatrix} a_{11} & a_{12} & a_{13} & a_{14} \\ a_{21} & a_{22} & a_{23} & a_{24} \\ a_{31} & a_{32} & a_{33} & a_{34} \\ a_{41} & a_{42} & a_{43} & a_{44} \end{pmatrix} \quad \text{and} \quad \vec{b} = \begin{pmatrix} b_1 \\ b_2 \\ b_3 \\ b_4 \end{pmatrix}.$$

We have 20 unknowns to determine. We will choose our matrix M and vector $\vec{b}$ so that $T(\vec{v}_1) = \vec{w}_1$, $T(\vec{v}_2) = \vec{w}_2$, and $T(\vec{v}_3) = \vec{w}_3$. This leads us to the following system of 12 equations:

$$T(\vec{v}_1) = \vec{w}_1 \implies \begin{cases} a_{11} + b_1 = 1, \\ a_{21} + b_2 = 1, \\ a_{31} + b_3 = 2, \\ a_{41} + b_4 = 0; \end{cases}$$

$$T(\vec{v}_2) = \vec{w}_2 \quad \Longrightarrow \quad \begin{cases} 2a_{11} + a_{12} + 2a_{13} + a_{14} + b_1 = 1, \\ 2a_{21} + a_{22} + 2a_{23} + a_{24} + b_2 = 1, \\ 2a_{31} + a_{32} + 2a_{33} + a_{34} + b_3 = 2, \\ 2a_{41} + a_{42} + 2a_{43} + a_{44} + b_4 = 1; \end{cases}$$

$$T(\vec{v}_3) = \vec{w}_3 \quad \Longrightarrow \quad \begin{cases} 2a_{12} + a_{13} + 2a_{14} + b_1 = 1, \\ 2a_{22} + a_{23} + 2a_{24} + b_2 = 1, \\ 2a_{32} + a_{33} + 2a_{34} + b_3 = 2, \\ 2a_{42} + a_{43} + 2a_{44} + b_4 = 2. \end{cases}$$

We give one solution for M and $\vec{b}$:

$$T(\vec{v}) = \begin{pmatrix} 0 & 1 & 1 & 0 \\ 1 & 1 & 0 & 1 \\ 0 & 2 & 0 & 1 \\ 0 & 1 & 0 & 0 \end{pmatrix} \vec{v} + \begin{pmatrix} 1 \\ 0 \\ 2 \\ 0 \end{pmatrix}.$$

You can check that this affine transformation sends the *SET* $\{(1, 0, 0, 0), (2, 1, 2, 1), (0, 2, 1, 2)\}$ to the *SET* $\{(1, 1, 2, 0), (1, 1, 2, 1), (1, 1, 2, 2)\}$, in that order. (We spared you the gory details, but the 12 equations can be reduced to just 8 equations. Since there are more variables than equations, the system is *underdetermined*, so we expect multiple solutions.)

We are now ready to answer the motivating question for this section:

Question: How many affine transformations are there?

We interpret this question geometrically: we can *uniquely* determine the affine transformation from its action on a collection of five points in *free position*, i.e., no three are collinear, no four are coplanar, and no five are cohyperplanar. Affine transformations preserve this property: if a collection of points is in free position, then so are its images under an affine transformation.

Thus, an affine transformation is completely determined by the images of five points in free position. We choose five special points: $(0, 0, 0, 0)$, $(0, 0, 0, 1)$, $(0, 0, 1, 0)$, $(0, 1, 0, 0)$, and $(1, 0, 0, 0)$. So, how many different images are there? The answer is $81 \times 80 \times 78 \times 72 \times 54 = 1,965,150,720$, the same number we found in chapter 6 when we determined the number of ways to place cards in our display of the entire deck.

Finally, how does this generalize to n dimensions? The argument here can be modified to give us an answer. The number of symmetries of the n-dimensional deck is

$$3^n(3^n - 1)(3^n - 3)(3^n - 3^2) \cdots (3^n - 3^{n-1}).$$

We point out that this is the numerator of $h(n, n - 1)$ from chapter 6.

8.5 PRESERVING THE NUMBER OF DIFFERENT ATTRIBUTES OF A *SET*

The affine transformations treat all *SET*s the same because all *SET*s are affinely equivalent. But one of the special features of SET is that we perceive *SET*s differently depending on how many attributes are the same. We've seen in the n-dimensional version of the game that the number of *SET*s with k attributes the same depends on both n and k. This was the focus of much of chapters 6 and 7.

Furthermore, paying attention to the different kinds of *SET*s is important when playing the game. *SET*s with only one attribute different are the easiest to find for many players, but there are fewer of those *SET*s than any other type.

With this motivation, we ask the following question:

Question: How many affine transformations preserve the number of attributes that are the same for all *SET*s?

We'll use linear algebra to answer this question. We'll see that the affine transformations that preserve the varying number of attributes correspond precisely to the different ways we could have assigned coordinates to the cards. First, let's give examples of the type of affine transformations that work.

1. Permutations of the coordinates. There are 4! ways to reorder the coordinates.

   - For instance, given a card, we can reorder the attributes by permuting the last three coordinates in a 3-cycle. Using vectors, this gives

     $$(v_1, v_2, v_3, v_4) \mapsto (v_1, v_4, v_2, v_3).$$

     For example, suppose our initial card is 1 Green Empty Oval. We encode this as the vector $(1, 0, 0, 1)$, as usual. Then the 3-cycle on the last three coordinates maps this vector to $(1, 1, 0, 0)$, which corresponds to the card 1 Purple Empty Diamond.

2. Permutations of the expressions of an attribute. Each attribute has three values, so there are 3! ways to permute each one.

   - For instance, we could swap the numbers 0 and 2 in the second attribute. In cards, this corresponds to swapping the colors green and red. Returning to the vectors, we have

     $$(v_1, v_2, v_3, v_4) \mapsto (v_1, 2 - v_2, v_3, v_4).$$

     In our example, let's see what this operation does to the card 1 Purple Empty Diamond, which was the output of the previous operation. This time, we find $(1, 1, 0, 0) \mapsto (1, 1, 0, 0)$. This means that this operation fixes this card (and all purple cards, in fact).
   - Finally, concentrating on the first attribute, we could also create the 3-cycle $1 \mapsto 2 \mapsto 0$. In vector form,

     $$(v_1, v_2, v_3, v_4) \mapsto (v_1 + 2, v_2, v_3, v_4).$$

     Returning to our running example, the vector $(1, 1, 0, 0) \mapsto (0, 1, 0, 0)$, so the output is the card 3 Purple Empty Diamonds.

Figure 8.16. The affine transformation preserves the number of attributes that are the same. Each *SET* has one attribute the same and three different.

We can combine these three operations to produce one affine transformation:

$$(v_1, v_2, v_3, v_4) \mapsto (v_1 + 2, 2 - v_4, v_2, v_3).$$

What does this transformation do to a *SET*? Let's look at an example.

Take the *SET* with vector representation $\{(1, 0, 0, 1), (2, 0, 2, 0), (0, 0, 1, 2)\}$. Applying the above transformation to the three vectors gives

$$(1, 0, 0, 1) \mapsto (0, 1, 0, 0), \qquad (2, 0, 2, 0) \mapsto (1, 2, 0, 2),$$

$$(0, 0, 1, 2) \mapsto (2, 0, 0, 1).$$

In cards, this maps the *SET* on the left in figure 8.16 to the one on the right. (The first card is the running example we used in our description of the operations above.) Note that each *SET* has one attribute the same and three different—this transformation preserves the number of varying attributes.

What matrix M and vector $\vec{b}$ correspond to this transformation? You can verify that

$$M = \begin{pmatrix} 1 & 0 & 0 & 0 \\ 0 & 0 & 0 & -1 \\ 0 & 1 & 0 & 0 \\ 0 & 0 & 1 & 0 \end{pmatrix} \quad \text{and} \quad \vec{b} = \begin{pmatrix} 2 \\ 2 \\ 0 \\ 0 \end{pmatrix}$$

do the job.

It's clear that operations that permute the numbers in a given coordinate will affect every card the same way. This won't change the number of attributes that are the same in a given *SET*. Neither will permuting the coordinates themselves.

How many transformations do these operations account for? There are $4! = 24$ permutations of the coordinates themselves, and there are

$3! = 6$ ways to permute each of the 4 coordinates. So we get $4! \times 6^4 = 31,104$ legal affine transformations using these operations.

Are there any more? The answer is no. This will take some work, and we'll use several of the ideas developed in the last two sections to explain why.

1. First, suppose $T(\vec{v}) = M\vec{v} + \vec{b}$ is an affine transformation that preserves the number of different attributes of all the *SET*s. Let's look at $T(0, 0, 0, 0)$ and $T(1, 0, 0, 0)$.

   Using $T(\vec{v}) = M\vec{v} + \vec{b}$, we get $T(0, 0, 0, 0) = \vec{b}$ since $M\vec{0} = \vec{0}$. Now $T(1, 0, 0, 0) = \vec{c}_1 + \vec{b}$, where $\vec{c}_1$ is the first column of the matrix M. To see why, just multiply a matrix M by the vector $(1, 0, 0, 0)$. This follows immediately from how we define the multiplication of a matrix and a vector.

2. Next, note that the *SET* $\{(0, 0, 0, 0), (1, 0, 0, 0), (2, 0, 0, 0)\}$ differs in just one attribute. That means the *SET* $\{T(0, 0, 0, 0), T(1, 0, 0, 0), T(2, 0, 0, 0)\}$ also differs in just one attribute. In particular, it means

   $$T(1, 0, 0, 0) - T(0, 0, 0, 0) \text{ has only one nonzero entry.}$$

   This nonzero entry will appear in the position corresponding to the one attribute of the image *SET* that is different.

3. But $T(1, 0, 0, 0) - T(0, 0, 0, 0) = \vec{c}_1$, the first column of the matrix M. We conclude from part (2) that $\vec{c}_1$ has only one nonzero entry.

4. The same thing is true for the second column—use the *SET*

   $$\{(0, 0, 0, 0), (0, 1, 0, 0), (0, 2, 0, 0)\}.$$

   This argument can be repeated for the third and fourth columns, too. So each column of the matrix has exactly one nonzero entry. This means that the entire matrix has only four nonzero entries.

5. Finally, each *row* of the matrix also has exactly one nonzero entry. Why? If this were false, then we would have a matrix with a row of 0s, which would force the transformation to fail to be one-to-one. (The real problem here is that such transformations

reduce dimension—they are not permutations of the cards in the deck.)

We conclude that the matrix M is a *signed permutation matrix*. This matrix has exactly one nonzero entry in each row and each column, and each nonzero entry is ± 1. Then there are 4! ways to pick the positions for the nonzero entries of M, and 2^4 ways to assign 1 or -1 to these positions. This gives $2^4 \times 4!$ signed permutation matrices.

But we still need to choose a translation vector $\vec{b}$. There are 3^4 vectors to choose from. It should be clear that translation will never change the number of varying attributes in a *SET*.

Putting all of this together tells us that the number of permutations that preserve the number of different attributes is $2^4 \times 4! \times 3^4$. But this equals 31,104—our two answers are the same! So linear algebra tells us we can't do anything except for the (relatively) simple operations of permuting the coordinates and permuting the numbers appearing in a fixed coordinate.

The general argument for the n-dimensional version of the game is identical. We conclude that there are $6^n \times n!$ permutations that preserve the number of varying attributes of the *SET*s in the n-dimensional game. And we needed linear algebra to prove there are no more.

***Executive Summary*:** Affine transformations map *SET*s to *SET*s. If you also want an affine transformation that preserves the number of varying attributes in all the *SET*s, then your transformation must be a composition of permutations of the coordinates and permutations of the numbers in a coordinate. There are $6^n \times n!$ such transformations.

We conclude this section with three pithy comments:

1. In the spirit of coding theory, call the number of attributes that are different in a given *SET* the *weight* of the *SET*. We've just determined when a transformation will preserve the weight of every *SET*. But our general proof used only the property that *SET*s of weight 1 are preserved. So we have a stronger conclusion:

    -   If an affine transformation preserves the weights of all *SET*s of weight 1, then it preserves the weights of all *SET*s.

2. Throughout this book, we've mentioned the arbitrary nature of our coordinate scheme. How many different schemes are possible, with the restriction that each coordinate will be 0, 1, or 2, and we will always work mod 3? The answer is $6^4 \times 4!$ for SET, and $6^n \times n!$ in general. To see this, note that any relabeling involves just the operations discussed in this section: a permutation of the n coordinates and n separate permutations of $\{0, 1, 2\}$ for each coordinate. But that's precisely what the affine transformations in this section do.

3. (*if you know some group theory*) In fact, we can describe the structure of the subgroup of weight-preserving transformations: $(S_3)^n \rtimes S_n$, where S_n is the symmetric group of all permutations of n symbols and $\rtimes$ represents a *semidirect product*.

EXERCISES

EXERCISE 8.1. For the *SET* and additional card in figure 8.3, verify that the procedure in section 8.2.2 does not depend on which card from the *SET* we choose to create the vector $\vec{w}$. That is, show that we get the same parallel *SET* when we use the second or third card in the *SET* to obtain $\vec{w}$.

EXERCISE 8.2. This exercise justifies using direction vectors to determine whether two *SET*s are parallel.

a. Suppose you are given a *SET* S. Show that the direction vector is uniquely determined up to multiplication by 2, i.e., show that there is a nonzero vector $\vec{d}$ such that, given any pair of vectors in the *SET* $\vec{u}$ and $\vec{v}$, either $\vec{u} - \vec{v} = \vec{d}$ or $\vec{u} - \vec{v} = 2\vec{d}$.

b. Use part (a) to show that two *SET*s are parallel if and only if their direction vectors are multiples of each other.

EXERCISE 8.3. Define a relation on the class of all $3^{n-1}(3^n - 1)/2$ *SET*s in n-dimensional SET as follows: two *SET*s S_1 and S_2 are related if $S_1 = S_2$ or they are parallel. Use the vector formulation of parallelism to show that this relation is *transitive*, i.e., show that if S_1 is parallel to S_2 and if S_2 is parallel to S_3, then S_1 is parallel to S_3.

TABLE 8.6.

Create a plane using the three non-collinear points $\vec{x}$, $\vec{y}$, and $\vec{z}$.

$\vec{x}$	$\vec{y}$	
$\vec{z}$		

EXERCISE 8.4. Look at figure 8.6, the entire deck laid out nicely. Consider the card in the upper left, 2 Green Striped Squiggles.

a. Start with a *SET* containing that card that lies in the subplane in the upper left (other than the one described in the chapter). Identify (and describe) the 26 *SET*s parallel to your *SET*.

b. Now take a *SET* containing that card that doesn't lie entirely in the subplane in the upper left (other than the one described in the figure). Identify (and describe) the 26 *SET*s parallel to your *SET*. Bonus points for a really spread-out *SET*.

EXERCISE 8.5. Choose three cards that do not form a *SET*, with corresponding vectors $\vec{x}$, $\vec{y}$, and $\vec{z}$.

a. Fill in the spots for the remaining positions in table 8.6 to form a plane. Note: All of your answers should be written in terms of $\vec{x}$, $\vec{y}$, and $\vec{z}$.

b. Show that the labelings in table 8.6 and table 8.2 are equivalent by finding equations relating the four vectors $\vec{v}_1$, $\vec{v}_2$, $\vec{v}_3$, and $\vec{w}$ from table 8.2 to the three vectors $\vec{x}$, $\vec{y}$, and $\vec{z}$ here.

EXERCISE 8.6. Use vectors to show that two disjoint *SET*s are parallel if and only if there is a plane that contains them both.

EXERCISE 8.7. Pick your favorite card, and find two *SET*s containing it, where each *SET* has one attribute that is the same. For this exercise, make sure the two *SET*s are different in which attribute is the same.

a. Construct the plane containing those two *SET*s. Show that every *SET* in this plane has exactly one attribute the same.

b. The plane has four classes of parallel *SET*s, and these partition the 12 *SET*s in your plane. Note that the attribute that's the same is different for each of these classes, i.e., one class will be the same in color, one class in number, one in shading, and one in shape (where all of the other attributes will vary for the given class). Use vectors to explain why this must be true.

c. Show that every card in the deck that's not in your plane differs from exactly one of the cards in the plane in exactly one attribute. Take each card in the plane you made in (a), and collect all eight cards that are different from that card in exactly one attribute. (We call these *code planes*. For more information about this plane, see "Error detection and correction using SET", in *The Mathematics of Various Entertaining Subjects: Research in Recreational Math*, J. Beineke and J. Rosenhouse eds., Princeton University Press, 2015.)

EXERCISE 8.8. Suppose T is an affine transformation. Use the fact that T takes *SET*s to *SET*s to show that

a. T preserves intersets;
b. T preserves planes;
c. T preserves hyperplanes;
d. T preserves collections of cards (of the same size) that contain no *SET*s.

EXERCISE 8.9. Find the unique affine transformation that sends all the following vectors to the given images:

$$(1, 0, 0, 0) \mapsto (1, 1, 2, 0),$$

$$(2, 1, 2, 1) \mapsto (1, 1, 2, 1),$$

$$(0, 1, 0, 2) \mapsto (2, 0, 1, 0),$$

$$(2, 2, 2, 0) \mapsto (2, 0, 0, 1),$$

$$(1, 1, 1, 2) \mapsto (0, 0, 0, 0).$$

EXERCISE 8.10. Show that affine transformations preserve parallelism, i.e., if T is an affine transformation and A and B are parallel *SET*s, then so are $T(A)$ and $T(B)$.

PROJECTS

PROJECT 8.1. (Perpendicular *SET*s) We can define *perpendicular SETs*. First, we need to define the dot product of two vectors:

Let $\vec{v}_1 = (a_1, b_1, c_1, d_1)$ and $\vec{v}_2 = (a_2, b_2, c_2, d_2)$ be two vectors. Define the *dot product* (also called the *inner product* or *scalar product*) as

$$\vec{v}_1 \cdot \vec{v}_2 = a_1 a_2 + b_1 b_2 + c_1 c_2 + d_1 d_2 \pmod{3}.$$

Figure 8.17. Two perpendicular *SET*s. The direction vectors satisfy $\vec{d}_1 \cdot \vec{d}_2 = 0$ (mod 3).

In ordinary Euclidean geometry, the dot product equals the cosine of the angle θ between the two vectors times the lengths of the vectors:

$$\vec{u} \cdot \vec{v} = |\vec{u}||\vec{v}| \cos\theta.$$

Since $\cos 90° = 0$, we immediately get the following way to tell if two vectors are perpendicular:

$$\vec{u} \perp \vec{v} \quad \text{if and only if} \quad \vec{u} \cdot \vec{v} = 0.$$

To make this work for SET, we will use direction vectors for *SET*s. As before, we compute the direction vector $\vec{d}$ by taking the difference of any two vectors in the *SET*. Consider the two *SET*s in figure 8.17.
Then we compute the direction vectors:

$$SET\ 1: \vec{d}_1 = (2, 1, 2, 1) - (1, 0, 2, 2) = (1, 1, 0, 2),$$

$$SET\ 2: \vec{d}_2 = (2, 1, 1, 0) - (2, 0, 0, 2) = (0, 1, 1, 1).$$

Then $\vec{d}_1 \cdot \vec{d}_2 = (1, 1, 0, 2) \cdot (0, 1, 1, 1) = 1 \times 0 + 1 \times 1 + 0 \times 1 + 2 \times 1 = 0$ (mod 3), so these *SET*s are perpendicular. Note that using $\vec{d}_2 = (2, 2, 2, 1) - (2, 0, 0, 2) = (0, 2, 2, 2) = 2(0, 1, 1, 1)$ makes no difference: $(1, 1, 0, 2) \cdot (0, 2, 2, 2) = 0$ (mod 3), too.

a. Find all *SET*s that are perpendicular to themselves. (Yep—that can happen.)
b. Show that perpendicularity is not preserved by affine transformations, i.e., find two perpendicular *SET*s A and B and a transformation T with $T(A)$ and $T(B)$ not perpendicular.
c. Show that a transformation T preserves perpendicularity if and only if T is one of the special transformations of section 8.5.
d. How many *SET*s are perpendicular to a given *SET*? Show that this number does not depend on the *SET* chosen, and describe the geometric structure of all the *SET*s perpendicular to a given *SET*.

Figure 8.18. Two perpendicular *SET*s intersecting at a card.

e. Suppose A, B, and C are three *SET*s. Show that if A and B are perpendicular *SET*s, and A and C are parallel, then B and C must be perpendicular.

f. It seems difficult[7] to determine if two *SET*s are perpendicular at a glance. To simplify this, we'll concentrate on intersecting *SET*s. Consider the *SET*s in figure 8.18. Show that if a *SET* A is perpendicular to both of these *SET*s, then A is perpendicular to every *SET* in the plane determined by these *SET*s.

g. Find any characterization of perpendicularity that could be quickly checked without using coordinates. (Try for something in the spirit of the cyclic attributes property that parallel *SET*s enjoy.)

PROJECT 8.2. (Affine closure) We've seen that planes and hyperplanes are *closed* in the following sense: given any two cards in a plane or hyperplane, the third card that completes the *SET* determined by the first two cards is in the plane or hyperplane. This seems to be the reason the Set Enterprises website refers to planes as "magic squares."

You may have seen closure before: it appears in algebra, where you are often interested in the subsets of an algebraic object (a group, ring, vector space,...) that are closed under binary operations. They also appear in topology, where a subset is closed if its complement is open.

We are interested in *affine closure*. We need a few definitions:

- A point $P \in AG(n, 3)$ is an *affine combination* of the collection of points $\{P_1, P_2, \ldots, P_k\}$ if both of the following conditions hold:

  a. $P = c_1 P_1 + c_2 P_2 + \cdots + c_k P_k$, where each $c_i = 0$, 1, or 2, and
  b. $c_1 + c_2 + \cdots + c_k = 1 \pmod 3$.

- A subset $S \subseteq AG(n, 3)$ is *affinely closed* if every affine combination of points of S is in S.

---

[7] Honestly, finding a pattern here looks hopeless. But there's hope.

- If $S \subseteq AG(n, 3)$, we define $\overline{S}$ to be the smallest affinely closed subset containing S. We call $\overline{S}$ the *affine closure of* S.

With this background, do the following:

a. Suppose $S = \{P_1, P_2\}$ is any collection of two SET cards. Show that $\overline{S}$ is the *SET* containing those two cards.

b. Show that a subset of $AG(4, 3)$ is affinely closed if and only if it is a point, a *SET*, a plane, a hyperplane, or the entire deck.

c. Let $S \subseteq AG(n, 3)$ be any subset of points. Show

   i. $S \subseteq \overline{S}$, and
  ii. $\overline{\overline{S}} = \overline{S}$.

d. Show that we can build $\overline{S}$ from S by repeatedly adding points to S as follows:

   i. Replace S by $S \cup S'$, where S' is the collection of all points that complete *SET*s with pairs of points of S.
  ii. Repeat step (i) until no new points are added.

Apply this procedure to the following subsets of the SET deck, keeping track of the number of times step (i) needs to be applied before you reach $\overline{S}$:

- S consists of two cards;
- S consists of three cards that do not form a *SET*;
- S consists of four cards that are not coplanar;
- S consists of five cards that are not cohyperplanar.

e. What happens if you remove the condition $c_1 + c_2 + \cdots + c_k = 1$ given above?

Affine Geometry Plus

9.1 INTRODUCTION

In chapter 5, we explored the connections between SET and affine geometry. We revisit that connection here and introduce some new geometric concepts. Recall that each card is a point in our geometry, and each *SET* is a line. The entire deck forms AG(4, 3), the affine geometry of dimension 4 and order 3. In this context, *dimension* 4 corresponds to the number of attributes in the game, and *order* 3 corresponds to the fact that each line contains exactly three points, i.e., three cards make a *SET*.

Using this connection, you can also visualize the finite geometries AG(3, 3) and AG(2, 2) with SET cards. For AG(3, 3), we can take the 27 red cards, for example, where the three varying attributes are number, shading, and shape; these cards form a hyperplane. Similarly, the nine red solid cards form a plane—these cards correspond to AG(2, 3). As a reminder, figure 9.1 shows all the lines in AG(2, 3).

In this chapter, we concentrate on a fundamental question in finite geometry that predates the invention of the game by several decades and which is getting a great deal of attention in the mathematical community as this book goes to press:

- What is the maximum number of points you can have with no three on a line?

From the viewpoint of SET, this question becomes,

- What is the maximum number of cards that do not include a *SET*?

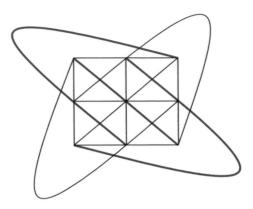

Figure 9.1. The finite geometry AG(2, 3) is an affine plane.

TABLE 9.1.
The largest cap sizes in dimensions ≤ 6.

Dimension	1	2	3	4	5	6
Cap size	2	4	9	20	45	112

Geometers call a collection of points in AG(n, 3) that contains no lines a *cap*. The largest possible size of a cap[1] is known in AG(n, 3) only for $n \leq 6$.

The game of SET corresponds to dimension 4 in table 9.1. Thus, the maximum number of cards that do not contain any *SET*s is 20. You can find this fact discussed in many places on the web, along with various proofs. There's also an excellent proof in B. Davis and D. Maclagan's wonderful article "The card game SET" (*Mathematical Intelligencer* 25, no. 3 (2003), 33–40). We can also interpret the results for $n = 1$, 2, and 3 by looking for caps in planes and hyperplanes; this will help build intuition for $n = 4$. Using the cards to visualize the caps, we'll see that the maximum-size caps have some interesting geometric structure. One goal in this chapter is to analyze this structure. We'll also look at a variation of SET that is based on projective geometry, rather than affine.

One final comment for now is that the connection between geometry and SET goes in both directions: using the cards allows us to visualize and understand the geometry in new ways, and the (very

---

[1] Large cap sizes may spark pictures of big heads, or perhaps sinking ships. We leave other puns to the interested reader.

well-developed) geometric theory gives rise to facts about collections of cards and *SET*s in SET. We will use both directions, advancing our understanding of both geometry and the game.

9.2 MAXIMAL CAPS

We begin with a comment on terminology. One of the difficulties in mathematics is that people who come up with a "new" mathematical idea need to name it. This means we cannot always be certain that a concept is, in fact, new: someone else might have come up with the same idea earlier, but called it something else. Who would guess that the term "cap" would be used for the collections of cards we're interested in?[2]

Another difficulty is that the standard term for the largest collections of points containing no lines is "maximal cap," despite the fact that mathematicians would usually call such a thing a *maximum* cap. A *SET* having some property is usually called *maximal* with respect to a property if adding additional elements to the collection destroys the property, though it might not be the largest such object. However, since this name is now standard, we will stick with it. We will use the following terminology: a "complete cap" is one that may not be of the largest possible size, but where adding any other point creates a line, while a "maximal cap" is a (complete) cap of largest possible size. We'll use these terms throughout the chapter.

While every maximal cap is a complete cap, in AG(3, 3) and higher dimensions, you can have complete caps that are not maximal. For example, consider the eight solid cards pictured in figure 9.2. Those cards form a complete cap in AG(3, 3): you can verify that every solid card completes at least one *SET* with two cards in the cap. However, the size of a maximal cap is nine cards, as indicated in table 9.1.

People have been interested in finding the maximal caps in various finite geometries since at least the 1940s; it's a bit hard to be certain because the terminology has changed. As far as we can tell, R. C. Bose ("Mathematical theory of the symmetrical factorial design," *Sankhyā: The Indian Journal of Statistics* **8** (1947), 107–166) was the first to

---

[2] No one.

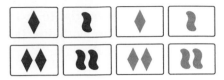

Figure 9.2. A complete cap in AG(2, 3): any solid card completes at least one *SET* with two cards from this collection of cards.

explicitly enumerate maximum-size line-free collections of points in AG(3, 3); he showed that the answer is 9, as table 9.1 indicates. In 1970, G. Pellegrino ("Sul massimo ordine delle calotte in $S_{4,3}$" [The maximal order of the spherical cap in $S_{4,3}$], *Matematiche (Catania)* **25** (1970), 149–157) proved that the answer is 20 in AG(4, 3). This paper is written in Italian and has not been translated into English.[3]

For dimensions 2, 3, and 4, we will be interested in several questions related to maximal caps:

1. How many cards are there in a maximal cap?
2. How many different maximal caps are there?
3. How many different maximal caps are there, *up to affine equivalence*? (See section 8.4.)
4. What is the geometric structure of a maximal cap?
5. What are the sizes of all the complete caps?

We will answer some of these questions in the chapter, and leave some as exercises and projects.

9.2.1 Caps in Dimension 2

Let's start small. In figure 9.3, you'll find a nice plane containing all the cards with 2 solid symbols. What's the largest collection of cards you can make with no *SET*s? Can you find a smaller collection of cards with no *SET*s?

It's not hard to see that any collection of three cards that aren't a *SET* can be augmented without creating a *SET*. Furthermore, it's not hard to see that *any* four cards that don't contain a *SET* will form a maximal cap. Jordan Awan wrote the Cap Builder applet

---

[3] We haven't read it. Ci scusiamo!

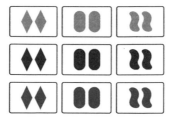

Figure 9.3. A nice plane.

●	●	1
●	●	1
1	1	2

Figure 9.4. The Cap Builder shows a maximal cap in two dimensions. The numbers indicate the number of lines each of the five remaining points completes with the cap.

http://webbox.lafayette.edu/~mcmahone/capbuilder.html, which allows you to choose points to form a cap in any dimension between 2 and 7.

Figure 9.4 contains output from this program. If we want our cap to be the four cards in the upper left of the plane in figure 9.3 (2 Green Solid Diamonds, 2 Green Solid Ovals, 2 Purple Solid Diamonds, 2 Purple Solid Ovals), select those positions, which will be represented by 4 large black dots.

What do the red numbers in figure 9.4 tell us? Note that the number 1 appears in the top right position (corresponding to 2 Green Solid Squiggles); this means that adding this card to the cap will produce exactly 1 *SET*: the *SET* determined by the top row of figure 9.3. The 2 in the lower right of the figure indicates that the card 2 Red Solid Squiggles completes 2 different *SET*s with the cards from the cap.

Thus, this maximal cap is an *interset*: two lines through a single point, with that point removed; further, every maximal cap in two dimensions is an interset. We will call the point that completes two lines an *anchor point*. (In chapter 2, we called this point the *center* of the interset, but that has a different meaning in geometry, so we won't use it here.) In figure 9.4, the anchor point is in the lower right: 2 Red

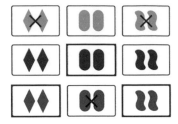

Figure 9.5. Choose the three boxed cards. Each of the cards not crossed out can be added to those three to make a maximal cap.

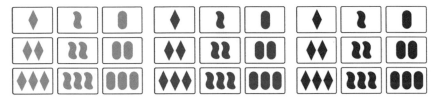

Figure 9.6. The affine geometry AG(3, 3) as a hyperplane in SET.

Solid Squiggles. In exercises 2.4 and 4.6, you had opportunities to show that the anchor point is unique.

Are there any complete caps that aren't maximal in AG(2, 3)? No. You can prove this by taking three cards that don't make a *SET*. Consider pairs of cards from those three and complete the *SET* for each pair. This eliminates three cards (shown with ✕'s on them in figure 9.5): these cards cannot be added to the cap. But then any of the three remaining cards can be added to the three original cards to make a maximal cap.

9.2.2 Caps in Dimension 3

Since AG(3, 3) is a hyperplane in AG(4, 3), we can use any hyperplane in the deck for our geometry. We'll use all of the solid cards as our model here. See figure 9.6.

In contrast to the situation in AG(2, 3), there are caps in AG(3, 3) that are complete but not maximal. Those caps have size 8, and you can explore them in project 9.1.

Not all complete caps with 8 points are affinely equivalent. Two different caps are pictured in figure 9.7. Note that the numbers in red

2	●	●	1	●	1	1	1	●
1	1	3	1	3	1	2	2	2
●	1	1	2	●	●	●	1	1

●	1	1	●	1	1	2	●	●
1	4	1	2	1	2	2	1	2
1	1	●	●	●	2	1	1	●

Figure 9.7. Two different complete caps in three dimensions.

●	2	●	2	2	2	●	2	●
●	2	2	●	2	2	2	●	●
2	2	●	2	2	2	2	2	2

Figure 9.8. A maximal cap in three dimensions.

are different for these caps: in the first instance, two points complete three lines, but for the second cap, no points complete three lines. Since affine transformations send lines to lines, a point that completes k lines with points from the cap will have to be sent to a point that completes k lines with the image of the cap. This tells us that these two caps cannot be transformed into one another with an affine transformation.

In AG(3, 3), all maximal caps contain nine points. One such cap is shown in figure 9.8. Note that every point not in the cap completes exactly two lines with points from the cap. Further, you can show that the sum of the coordinates of all the points in a maximal cap in AG(3, 3) is $\vec{0}$.

We saw that any collection of four points, with no three on a line, forms a maximal cap in AG(2, 3). Further, all of these four-point caps are affinely equivalent. It turns out this is also true in three dimensions:

• All maximal caps in AG(3, 3) are affinely equivalent.

This was originally proved by Bose using fairly sophisticated arguments, although it's also possible to construct a direct proof. But the details are too technical for us to worry about here.

●	●	1	●	●	1	1	1	2
●	●	1	●	●	1	1	1	2
1	1	2	1	1	2	2	2	4
●	●	1	●	●	1	1	1	2
●	●	1	●	●	1	1	1	2
1	1	2	1	1	2	2	2	4
1	1	2	1	1	2	2	2	4
1	1	2	1	1	2	2	2	4
2	2	4	2	2	4	4	4	8

Figure 9.9. A complete cap in four dimensions.

9.2.3 Caps in Dimension 4, the Game of SET

Now, we're ready for the maximal caps in AG(4, 3), i.e., collections of cards from the usual SET deck that are as large as possible, but contain no *SET*s. Since the largest possible cap has 20 cards, we now know that any collection of 21 cards is guaranteed to contain a *SET*. It's possible to find complete caps with 16, 17, 18, or 20 cards, but not 19 (any cap with 19 points is actually a subset of a maximal cap).

There's quite a large number of complete caps that are not affinely equivalent. If you go to the Cap Builder, there's a button that allows you to create a random complete cap. Playing with that for a while should convince you of two things: we weren't kidding[4] when we said that there are lots of complete caps that aren't affinely equivalent, and it's extremely rare that you get a maximal cap.

As before, it is useful to picture caps by using a grid to represent AG(4, 3), i.e., the entire deck. Here, we use a 9 × 9 grid, divided into 3 × 3 subgrids, to represent the 81 cards in the deck. Figure 9.9 gives an example of a complete cap of size 16.

Given how we can organize the cards, you can see that this cap can be created by taking all the cards that have

---

[4] We leave to the astute reader the task of finding the places where we were kidding. There are a bunch.

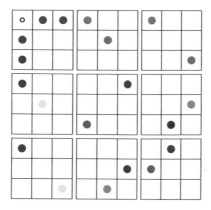

Figure 9.10. A maximal cap in AG(4, 3).

- number: 1 or 2 symbols;
- color: red or green symbols;
- shading: empty or striped symbols;
- shape: diamonds or ovals.

If we add any of the remaining 65 cards to this collection, we are forced to create a *SET*, so this is a complete cap. But this is not a maximal cap; those have 20 cards.

We are interested in the geometric structure of the 20 cards that form a maximal cap. It turns out the cap will consist of 10 lines through a point,[5] with that point then removed. If you're keeping track, that gives us 20 points, as advertised.

As we did in the two-dimensional case, we call the point on these 10 lines the *anchor* point of the cap. One example of a maximal cap is shown in figure 9.10, where the anchor point is in the upper left corner, and a pair of points in the cap are the same color if that pair forms a *SET* with the anchor point.

What does this cap look like in cards from the deck? We give an example in figure 9.11. The anchor card is 1 Green Empty Diamond, and it's outlined. It's a useful exercise to find all 10 *SET*s that contain the anchor card.

What else can be said about maximal caps in AG(4, 3)?

---

[5] Such a collection of lines is properly called a *pencil* of lines, which should remind the reader of group terms like *pride of lions* or *murder of crows*. We believe this is an excellent name.

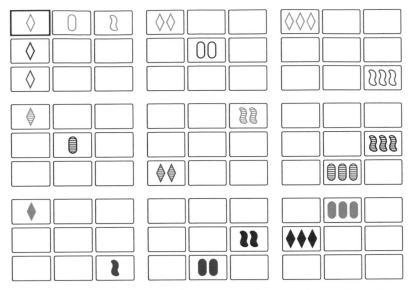

Figure 9.11. Four-dimensional complete cap, in cards; the anchor card is outlined in the upper left.

- As was the case in two and three dimensions, it turns out that all maximal caps in AG(4, 3) are affinely equivalent. This fact was first shown by R. Hill in 1983 ("On Pellegrino's 20-caps in $S_{4,3}$," *Annals of Discrete Mathematics* **18** (1983), 433–447).
- If you put the cap shown in figure 9.10 into the Cap Builder, you'll find that any point not in the cap (and not the anchor point) completes exactly three lines with points from the cap. A picture using the Cap Builder appears in figure 9.12; an example of a card not in the *SET* and the three *SET*s it completes (with the cards in figure 9.11) is shown in figure 9.13.
- Two maximal caps with different anchors must intersect. This was verified with a computer search by M. Follett, K. Kalail, E. McMahon, C. Pelland, and R. Won ("Partitions of AG(4, 3) into maximal caps," *Discrete Mathematics* **337** (2014), 1–8), and it was proved directly (without the computer) by J. Awan, C. Frechette, and Y. Li.
- It is possible to find disjoint caps that use the same anchor point. This will lead us to some beautiful partitions of the entire deck

A	●	●	●	3	3	●	3	3
●	3	3	3	●	3	3	3	3
●	3	3	3	3	3	3	3	●
●	3	3	3	3	●	3	3	3
3	●	3	3	3	3	3	3	●
3	3	3	●	3	3	3	●	3
●	3	3	3	3	3	3	●	3
3	3	3	3	3	●	●	3	3
3	3	●	3	●	3	3	3	3

Figure 9.12. A maximal cap of 20 cards, shown using the Cap Builder program. Note that each non-anchor point not in the cap completes exactly three lines with points in the cap. The point labeled A is the anchor—it's not in the cap, and it completes 10 *SET*s with the points in the cap.

Figure 9.13. The top card is not in the cap shown in figure 9.11. This card forms a *SET* with three pairs of cards in the cap.

into disjoint maximal caps. We study these in some detail in section 9.3.

9.2.4 Caps in Higher Dimensions

Table 9.2 presents what is known about maximal caps in dimensions 5 and 6. ("Lines completed" refers to how many lines every (non-anchor) point not in the cap completes with pairs of points in the cap.)

In five dimensions, there is no anchor point, but the coordinates of the points of the cap sum to $\vec{0}$ (as was the case in three dimensions). In six dimensions, there is an anchor point (as was the case in two and four dimensions); the cap consists of 56 lines through the anchor point, with that point removed. Notice that the maximal caps in even dimensions (2, 4, and 6) contain anchor points, while in nontrivial odd dimensions (3 and 5), there are no anchor points.

Table 9.2.
Known sizes of maximal caps.

Dimension	Size of max cap	All affinely equivalent?	Lines completed	Anchor point?
5	45	yes	5	no
6	112	yes	10	yes

What is known in dimensions greater than 6? Very little. You may find it surprising that no other maximal cap sizes are known. The results for dimension 5 were established in 2002 by Y. Edel, S. Ferret, I. Landjev, and L. Storme ("The classification of the largest caps in AG(5, 3)," *Journal of Combinatorial Theory A* **99**, no. 1 (2002), 95–110), and dimension 6 was completed by A. Potechin ("Maximal caps in AG(6, 3)," *Designs, Codes and Cryptography* **46**, no. 3 (2008), 243–259) when he was an undergraduate student in 2008. It would be great if a reader of this book were to find the size of a maximal cap in AG(7, 3).

Finally, this problem is not solely of interest to SET enthusiasts. In 2007, Fields Medalist Terence Tao wrote in his blog,

> Perhaps my favourite open question is the problem on the maximal size of a *cap set*. (https://terrytao.wordpress.com/2007/02/23/open-question -best-bounds-for-cap-sets/)

The Fields Medal is often referred to as the "Nobel Prize of Mathematics." Tao is one of the premier mathematicians of the twenty-first century, winning that prize in 2006. If nothing else, this should convince you that a general formula for the maximal cap size in n dimensions is probably hopeless.

Write cap(n) for the size of the largest cap in AG(n, 3). Currently, the best known bounds for cap(n) are

$$(2.2174\ldots)^n \leq \text{cap}(n) \leq c \cdot (2.756)^n$$

where c is a constant that does not depend on n. The lower bound is due to Edel, and the upper bound represents a major breakthrough, which was announced in May 2016 by Ellenberg and Gijswijt, building on previous work by Croot, Lev, and Pach.

Figure 9.14. A partition of AG(2, 3) into two maximal caps with anchor point A.

9.3 PARTITIONS OF MAXIMAL CAPS

Is it possible to break up the entire SET deck into maximal caps, all using the same anchor point, with no cards appearing in more than one cap? This is a question about partitioning a finite geometry into caps, and such questions have received a fair amount of attention from people who study finite geometry. Aided by the visualization provided by SET, we'll see how this is possible for $n = 4$.

That such a partition exists for the SET deck was first noticed by the mathematician Anthony Forbes of the UK (personal communication); since his discovery, two research groups have studied the geometric structure of these partitions. These groups have also shown that similar partitions are possible in lower dimensions.

9.3.1 Partitions in Dimension 2

In AG(2, 3), we know that there are exactly four lines through any point (corresponding to the four SETs that contain any card in figure 9.3). Start with a maximal cap in a plane, and call its anchor point A. The cap consists of two lines through A, with A removed. Then there are two more lines through A, and those two lines (again with A removed) also form a maximal cap. This means that we can use the four lines through A to decompose AG(2, 3) into two disjoint maximal caps, together with their common anchor point. Such a partition is shown in figure 9.14.

In figure 9.14, you can see the anchor point A along with the four lines through A. The four red points are a maximal cap and the four blue points are a disjoint maximal cap, so this gives us one partition. Similarly, the four solid points and the four open points are a second pair of disjoint maximal caps, so that is a different partition.

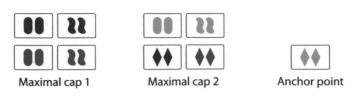

Maximal cap 1   Maximal cap 2   Anchor point

Figure 9.15. Two maximal caps plus their common anchor point partition AG(2, 3).

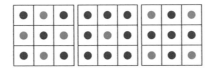

Figure 9.16. AG(3, 3) partitioned into three disjoint maximal caps.

Fixing the anchor point A, there are (at least) two natural counting questions:

- How many maximal caps are there that use anchor point A?
- How many partitions are there with anchor point A?

You'll get a chance to answer these questions yourself in exercise 9.2. In figure 9.15, you'll see the cards that correspond to one partition of AG(2, 3) into two disjoint maximal caps plus the anchor point (where we use the configuration of cards in figure 9.3). This partition corresponds to choosing the four red points for one cap and the four blue points for the other from figure 9.14.

Notice that, in AG(2, 3), once a maximal cap is chosen, the second cap in a partition is determined. In exercise 8.7(a), you were asked to prove that all intersets are affinely equivalent (this is true in any dimension), so all maximal caps in AG(2, 3) are affinely equivalent, so all partitions of AG(2, 3) are affinely equivalent. See exercise 9.3.

9.3.2 Partitions in Dimension 3

We know that AG(3, 3) consists of 27 points, and there are 9 points in a maximal cap. Are there partitions of AG(3, 3) into three disjoint maximal caps? Indeed there are! One such is pictured in figure 9.16. In the exercises, you'll be asked to prove that a given maximal cap is in a unique partition of AG(3, 3) into disjoint maximal caps, just as was the

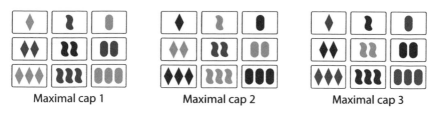

Figure 9.17. Partition of AG(3, 3) into three disjoint maximal caps.

case in AG(2, 3), i.e., once you've picked one maximal cap, the other two caps that complete the partition are completely determined.

In figure 9.17, you'll see the cards that correspond to this partition. Again, you can check that each cap contains no lines and that every solid card is in exactly one cap. There are other wonderful patterns in there, so you might want to spend a little time looking for them.

It's interesting to compare the three caps of figure 9.17 with the cards that form a plane (see figure 9.3). These collections of nine cards are opposite, in some sense: the nine cards that form a plane contain 12 *SET*s, while each of the three caps in figure 9.17 contains no *SET*s. You can explore more of the patterns in this partition in project 9.1.

9.3.3 Partitions in Dimension 4, the Game of SET

Are there partitions of AG(4, 3) into disjoint maximal caps? The numbers are promising: there are 81 points in total, 20 points in a maximal cap, and $81 = 4 \times 20 + 1$. There are 40 lines through a given point, and a maximal cap accounts for 10 such lines. In a perfect world, it would be possible to partition these 40 lines into four groups of 10 lines, with each collection of 10 lines a maximal cap (remembering that we never include the anchor point in the cap, of course).

It turns out that, at least in this instance, the world is perfect.[6] As we mentioned above, Anthony Forbes found such a partition in 2007. We give one example, both in grid form, and using the cards: figure 9.18 shows the partition using our grid and figure 9.19 shows the same partition in cards.

There is much more to explore with these partitions. Anthony Forbes also did a number of computer searches and found several

[6] Make up your own footnote here. The authors are speechless.

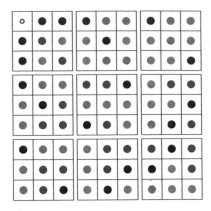

Figure 9.18. A partition of AG(4, 3).

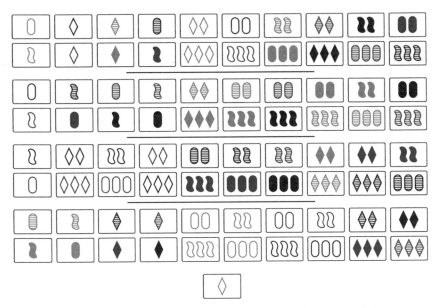

Figure 9.19. Partition of AG(4, 3) into four disjoint maximal caps, plus their common anchor point.

interesting facts:

- Given a particular maximal cap C, there are exactly 198 maximal caps disjoint from it, and each has the same anchor point.
- Every maximal cap is in 216 partitions, and every pair of disjoint maximal caps is in at least 1 partition.

Finally, two groups of undergraduate students proved the following:

- It is not the case that all partitions are affinely equivalent, but rather there are two affine equivalence classes. Work in understanding the geometric difference between these classes has given rise to an interesting substructure of the maximal caps. (M. Follett, K. Kalail, C. Pelland, R. Won, J. Awan, C. Frechette, and Y. Li)

9.3.4 Higher Dimensions

Are partitions like the ones we found in dimensions 2, 3, and 4 possible in dimension 5 or 6? Unfortunately, no. The maximal cap in AG(5, 3) has 45 points, and 45 does not divide 3^5. Similarly, 112 does not divide $3^6 - 1$, so this is impossible in AG(6, 3), too. However, there might still be some undiscovered structure these caps possess. And you might be the person who figures out what that structure is.

9.4 A PROJECTIVE VERSION OF SET

As we have mentioned (repeatedly), the game of SET is a model for a finite affine geometry. There are also finite projective geometries that can form the basis for SET-like games.

Davis and Maclagan's article from the *Mathematical Intelligencer* ("The card game SET," previously mentioned in this chapter) describes a projective version of SET. A game called Zero SumZ (or referred to as ProSET, short for Projective SET) was designed by A. Erickson, M. Guay-Paquet, and J. Lenchner; it can be played online at http://www .zerosumz.com and is available for purchase on the web. There are (at least) two other online versions: A. Geraschenko's at http://stacky .net/wiki/index.php?title=Projective_Set and D. Adams' at https://www .ocf.berkeley.edu/~dadams/proset/.

There are important differences between these games and SET. Some versions allow *SET*s with more than three cards, and these games usually require the symbols, or attributes, to appear an even number of times, which has a different feel from the game of SET.

246 • Chapter 9

There is another approach to modifying the game of SET to make it a projective geometry, taken by D. Burkholder. In a version called Complete SET, he starts with the actual SET deck and adds 40 cards to it to make a projective deck. An abstract of a presentation he gave on this game can be found online: https://jointmathematics meetings.org/amsmtgs/2168_abstracts/1106-a1-1254.pdf. While this version uses the regular SET cards (along with the additional cards), *SET*s in this version contain four cards, rather than three.

In the interest of completeness, we first give a brief introduction to projective geometry, then present another version of SET based on this geometry.

We begin by looking at the axioms for a projective plane. The main difference between affine and projective planes is that in an affine plane, the parallel postulate holds: given a line l, and a point P not on l, there is a unique line l' through P parallel to l. In the projective plane, there are no parallel lines at all; every pair of lines intersect.

Axioms for a Finite Projective Plane

Axiom 1. *There exist four points where no three are on the same line.*
Axiom 2. *Any two points lie on a unique line.*
Axiom 3. *Any two lines intersect in a unique point.*

There are several things to notice here. The first axiom rules out the trivial geometries where all the points lie on a line or where all lines pass through the same point. Note the symmetry between the second and third axioms. If we interchange the words "point" and "line" (and if we're willing to be flexible about the meaning of "lie on" and "intersect in"), then axiom 2 becomes axiom 3 and vice versa. This is called "point–line duality" for projective planes. Thus, once we prove the dual to the first axiom (namely that there are at least four lines, no three through the same point), then for any theorem that we prove, the words "point" and "line" can be interchanged, giving a new, dual theorem. For instance, suppose we have a projective plane with the property that every line has exactly three points. Then it must also be true that every point is on exactly three lines.

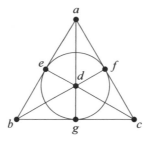

Figure 9.20. The Fano plane is a projective plane with three points on every line.

The simplicity of the axioms and the point–line duality make projective geometry a very attractive field. While projective geometry might seem less natural than affine geometry (because Euclidean geometry is affine), mathematicians have spent more time studying projective geometries. Moreover, standing on train tracks[7] should convince you that parallel lines appear to intersect "at infinity." In fact, much of the early work in projective geometry was devoted to understanding perspective in art.

Although the axiom systems are different, there is a close connection:

- Every affine plane is contained in a projective plane. A projective plane is obtained from an affine plane by adding a "line at infinity."

This connection has been exploited by mathematicians. In fact, most of what we know about caps in affine geometries comes from theorems about caps in projective geometries.

As usual, the best way to understand a new concept is by looking at examples. The Fano plane in figure 9.20 is the smallest projective plane, with seven points and seven lines (where the circle is also a three-point line). The first person to describe this geometry was G. Fano ("Sui postulati fondamentali della geometria proiettiva," *Giornale di Matematiche* **30** (1892), 106–132). This paper is in Italian, just as Pellegrino's was.[8] There is a long history of research on projective geometry in Italy, dating back to the Renaissance.

---

[7] Please do this only when no trains are nearby. Please.
[8] Should we feel guilty for not reading papers in Italian? Probably.

TABLE 9.3.
Assignment of coordinates to points in the Fano plane.

Point	a	b	c	d	e	f	g
Coordinates	$(1, 0, 0)$	$(0, 1, 0)$	$(0, 0, 1)$	$(1, 1, 1)$	$(1, 1, 0)$	$(1, 0, 1)$	$(0, 1, 1)$

There is a lot of symmetry in this geometry. Here are some relevant facts you might notice:

1. There are seven points and seven lines in the geometry. (Axiom 1 is satisfied.)
2. Every pair of points determines a unique line. (Axiom 2 is satisfied.)
3. Every pair of lines meets at a unique point, i.e., there are no parallel lines. (Axiom 3 is satisfied.)
4. Every point is on three lines.
5. Every line has three points.

To model a card game on the Fano plane, we could create a card for each of the seven points. Then the "*SET*s" could be the lines, exactly matching the situation for SET. While this game would not be interesting to play (there aren't enough cards), the geometry has the properties that make SET so attractive: every *SET* has three cards, and every pair of cards is in a unique *SET*. The projective version of the game we describe below is based on this geometry. To understand how this projective game works, we will need coordinates, much as we have throughout this book.

While the coordinate scheme we use here might seem like it's coming out of the blue, it's the standard approach to representing projective planes by vectors. We will still use modular arithmetic, but we'll now be working mod 2, instead of mod 3. We will use ordered triples (a, b, c), where each coordinate is 0 or 1, to assign coordinates to the seven points in the Fano plane, except that we won't assign the (0,0,0) vector to any of our points, leaving the seven ordered triples we need.

We give our assignment in table 9.3.

The Fano plane is denoted PG(2, 2), for projective geometry of dimension 2, where our arithmetic is done modulo 2. Working mod 2 gives these coordinates a remarkable (and quite familiar) property:

- Three points are collinear if and only if their sum is $(0, 0, 0)$ (mod 2).

For example, the three points a, b, and e are a line in the geometry. Adding their coordinates,

$$(1, 0, 0) + (0, 1, 0) + (1, 1, 0) = (2, 2, 0) = (0, 0, 0) \quad \text{(mod 2)}.$$

The other six lines also sum to $(0, 0, 0)$ (mod 2). Furthermore, if you look at three points that do not lie on a line, their coordinates will not sum to $(0, 0, 0)$ (mod 2). For example, the points a, c, and d are not collinear, and

$$(1, 0, 0) + (0, 0, 1) + (1, 1, 1) = (2, 1, 2) = (0, 1, 0) \quad \text{(mod 2)}.$$

The main point is that this matches the situation for SET, except that the arithmetic is done modulo 2. (From the viewpoint of linear algebra, three points are collinear if and only if the corresponding vectors are linearly dependent.)

The Fano plane has three points per line, with every pair of points determining a unique line. These are desirable properties, but, as we remarked above, this is too small to make an interesting game. In order to make this a game we could reasonably play, we need to increase the number of points. To do this, we'll add three more coordinates, so our points now correspond to ordered 6-tuples $(x_1, x_2, x_3, x_4, x_5, x_6)$, where each $x_i = 0$ or 1. We remove the 6-tuple $(0, 0, 0, 0, 0, 0)$, leaving us with a deck of $2^6 - 1 = 63$ cards. Although it is difficult to visualize this five-dimensional object, a model of the three-dimensional PG(3, 2) is shown in figure 9.21. In exercise 9.7, you are asked to show that there are a total of 651 lines, each of which will be a "*SET*" in the game.

How do you turn the coordinates into cards? Our version was inspired by a comment in the Davis and Maclagan paper; we chose it because we wanted to maintain the property of having three expressions for attributes. Group the coordinates in pairs, writing $(a_1, a_2; b_1, b_2; c_1, c_2)$, where each coordinate is 0 or 1 (and we insist that not all of these coordinates are chosen to be 0).

Figure 9.21. A model of three-dimensional projective space PG(3, 2) has three points per line. This geometry has 15 points and 35 lines.

Now we construct the deck by making faces on the cards; each pair of coordinates represents one attribute: eyes, mouth, hair. Our faces will now have eyes that are closed, brown, blue, or red,[9] a mouth that is missing, straight, smiling, or open, and a head that's bald, or has blonde, brown, or black hair. See table 9.4 for an assignment and figure 9.22 for a picture of some of the cards we made.

[9] Vampires?

TABLE 9.4.
Translating coordinates to cards.

Coordinates	Card	Coordinates	Card
$(0, 0; *, *; *, *)$	eyes closed	$(*, *; 0, 1; *, *)$	mouth smiling
$(1, 0; *, *; *, *)$	brown eyes	$(*, *; 1, 1; *, *)$	mouth open
$(0, 1; *, *; *, *)$	blue eyes	$(*, *; *, *; 0, 0)$	no hair
$(1, 1; *, *; *, *)$	red eyes	$(*, *; *, *; 1, 0)$	blonde hair
$(*, *; 0, 0; *, *)$	no mouth	$(*, *; *, *; 0, 1)$	brown hair
$(*, *; 1, 0; *, *)$	mouth closed	$(*, *; *, *; 1, 1)$	black hair

Figure 9.22. A Fano plane in cards.

A "*SET*" is a collection of three cards whose coordinates sum to the zero vector. Note that the card with all attributes missing (eyes closed, no mouth, no hair) corresponds to the vector $(0,0,0,0,0,0)$. This card is not in the deck; using our definition, it would never be in any *SET*.

How can you recognize a *SET*? There are three ways an attribute can appear or not in a *SET*:

1. The attribute could be missing on all three cards.
2. One expression for the attribute could appear on two cards and the third card would have that attribute missing.

3. Each card has a different one of the three visible expressions for the attribute.

For example, listing the cards in the order (eyes, mouth, hair), the three cards

(brown, none, black),   (brown, none, brown),   (closed, none, blonde)

form a *SET*, since this corresponds to the three vectors

$$(1, 0; 0, 0; 1, 1) + (1, 0; 0, 0; 0, 1)$$
$$+(0, 0; 0, 0; 1, 0) = (0, 0; 0, 0; 0, 0) \quad (\text{mod } 2).$$

In exercise 9.8, you are asked to verify that these really are the only ways that the cards will sum to the zero vector. Note that if you have three cards with the same visible expression of an attribute (for example, three with red eyes), the coordinates for that attribute could *not* sum to (0,0). So you'll never see the same attribute on each card in a *SET*.

We call this game PSET. The rules for PSET are more complicated than the rules for SET. There is also an unsettling lack of symmetry in the rules, since eyes closed, no mouth, and no hair are treated differently from the other expressions. Also, the fact that you can't have the same expression appearing three times makes good SET players worse at this game. See if you can find the seven "*SET*s" in figure 9.22. [Hint: Think about the lines in the Fano plane.]

Much of the analysis that we've done for SET can be done for PSET. For example, if you take all the cards with red eyes and blue eyes, you can't have any *SET*s, so you can have a cap with 32 cards! (This is the best possible: J. Bierbrauer and Y. Edel point out that 2^k is in fact the size of the unique maximal cap in PG(k, 2) in "Large caps in small spaces," *Designs, Codes and Cryptography* **23**, no. 2 (2001), 197–212.)

We have played this game. It has the potential to be a fun game, but the asymmetry in the rules makes it more difficult than SET. Like most things in life, (1) it would take a lot of practice to get good at it, but (2) you do get better at it as you play. And in that way, it is certainly similar to SET.

EXERCISES

EXERCISE 9.1. In a plane, take any three points that are not a *SET*. Show that there are exactly three maximal caps containing those three points. What can you say about the anchor points for each of those three maximal caps?

EXERCISE 9.2. These questions all refer to the plane AG(2, 3).

a. Choose a point A. How many maximal caps are there with anchor point A?
b. Again, with A chosen, how many partitions into two disjoint maximal caps are there with anchor point A?
c. How many maximal caps are there in all of AG(2, 3)?
d. How many partitions into two pairs of maximal caps together with their common anchor point are there in all of AG(2, 3)?

EXERCISE 9.3. Use linear algebra (chapter 8) to prove that any two maximal caps in AG(2, 3) are affinely equivalent, either by constructing an affine transformation that takes one to the other or by creating one by specifying the images of three points. (It's not illegal to do both.) Conclude that all partitions of AG(2, 3) into two maximal caps plus a point are affinely equivalent.

EXERCISE 9.4. Suppose C is a maximal cap in AG(3, 3). Show that if you slice AG(3, 3) by three parallel planes, then those planes can intersect C in one of only two ways: either each plane contains three points, or two planes contain four points and one contains one point.

EXERCISE 9.5. Prove that a given maximal cap in AG(3, 3) is in a unique partition of AG(3, 3) into mutually disjoint maximal caps. [Hint: Because any two maximal caps are affinely equivalent, it's enough to show the statement for a single maximal cap.]

EXERCISE 9.6.

a. Use the Cap Builder to find a maximal cap in AG(4, 3) that's different from the one shown in the chapter. It will help a great deal to remember that you'll need 10 lines through a given point.
b. Find a partition of AG(4, 3) that includes your maximal cap.

EXERCISE 9.7. The PSET deck is based on the geometry PG(5, 2).

a. Show that there are 651 lines.
b. Find the number of lines in PG(n, 2). If you've read section 6.4 on q-binomial coefficients, show that your answer is given by $\begin{bmatrix} n+1 \\ 2 \end{bmatrix}_2$.

EXERCISE 9.8.

a. Verify that the three ways of finding a *SET* in PSET (listed in the chapter) correspond to the coordinates summing to $(0, 0; 0, 0; 0, 0)$ (mod 2).
b. Verify that no other collections of three cards can sum to $(0, 0; 0, 0; 0, 0)$ (mod 2).
c. Verify that any pair of cards determines a unique *SET*.

EXERCISE 9.9. Make up your own deck of PSET cards, and find someone to play with. Who won?

PROJECTS

PROJECT 9.1. This project explores the complete caps that are in AG(3, 3).

a. Prove that no collection of seven points in AG(3, 3) can be a complete cap.
b. How many different complete caps of size 8 are there, up to affine equivalence, in AG(3, 3)? Prove your assertion.
c. In figure 9.17, we gave a partition of AG(3, 3) into three disjoint maximal caps. No *SET*s are contained within any cap (that's what it means to be a cap), but these 27 cards contain 117 *SET*s in total. Some of those *SET*s are composed of one card from each of the three caps, and the rest are composed of two cards from one cap and one from another. Thus, there are seven different kinds of *SET*s.

  i. Count the number of *SET*s of each type for the partition of figure 9.17.
  ii. Will your numbers be the same for any partition of AG(3, 3) into maximal caps?

PROJECT 9.2. After doing exercise 9.4, analyze what the configurations of cards in parallel plane slices must look like.

a. If you start with three points in one plane, what must the three points in the next plane look like? What about the last plane?

b. If you start with four points in one plane, what must the plane containing four points look like? What is the relation of the single point to those two planes?

PROJECT 9.3. Now that we've developed the game of PSET, you can redo many of the counts we've done for SET in that context.

a. How many *SET*s are there? How many *SET*s include a given point?
b. Describe the different kinds of *SET*s (recall, in ordinary SET, we had *SET*s with one attribute the same and three different, etc.), and then count how many *SET*s of each kind there are.
c. How many Fano planes are there?
d. How many intersets are there?
e. Make up one more good question and answer it.
f. What other questions can you explore with this game?

Computing and Simulations

10.1 INTRODUCTION

Throughout this book, we've used SET to generate interesting counting and probability questions.[1] We've answered several of these questions, primarily in chapters 2, 3, 6, and 7. But, along the way, we've also seen plenty of good questions that seem difficult or impossible to answer exactly. In this chapter, we'll revisit some of those questions using computer simulations to estimate various counts, probabilities, and expected values. Some of the results we present surprised us, and we believe these deserve further exploration.

Using computer simulations to estimate answers to difficult questions is usually called the *Monte Carlo method*, after the Monte Carlo Casino. The earliest Monte Carlo simulations were done in the 1940s for the Manhattan Project, which developed the first nuclear weapons. Since that time, this method has been an extremely powerful tool in the physical sciences and in mathematics, often giving precise estimates for otherwise intractable problems.

In 2004, during an NSF-funded undergraduate research summer at Lafayette College, David Eisenstat (then an undergraduate at the University of Rochester) wrote a computer program that simulated playing the game millions of times to estimate the probabilities for how many cards are left on the table at the end of the game. Two years later, Maureen Jackson, a Lafayette College student, ran more simulations while exploring SET in her honors thesis. Brian Lynch,

---

[1] Well, at least we found them interesting.

another Lafayette student studying the game in 2013, modified the program to run many additional simulations. In this chapter, we'll summarize some of these results. David and Brian kindly allowed us to share some of their Java code, which is available on the book's website; you can use this code as is or modify it to run some of your own simulations.

Here is a brief synopsis of how these programs work: First, encode the cards as integers from 0 to 80 as follows: if the card has (mod 3) coordinates (a, b, c, d), then assign this card the integer $27a + 9b + 3c + d$. For example, the card 2 Green Striped Ovals corresponds to the 4-tuple $(2, 0, 1, 1)$, which gives $27 \times 2 + 9 \times 0 + 3 \times 1 + 1 = 58$. (This corresponds to the base-3 expansion of $58 = 2011_3$.)

Next, choose a random permutation of the numbers from 0 to 80 to simulate shuffling the deck. The first 12 numbers in that permutation represent the first layout of cards on the (virtual) table. Once a *SET* is chosen, the three cards in the *SET* are removed and the next three cards are "dealt."

To "play" the game, start with the first 12 cards, and follow this algorithm:

1. There are 12 cards available (or fewer if the deck has run out). Search all triples of those cards and list all *SET*s found. If there is at least one *SET*, go to (2). If there are no *SET*s, and there are cards left in the deck, add three cards and go to (3). If there are no *SET*s, and there are no cards left in the deck, go to (4).

2. Choose one *SET* from the list of *SET*s (how that choice is made will be explored later in the chapter) and remove it. If there are cards left in the deck, add three additional cards, and return to (1). If there are no cards left in the deck, there's nothing to add, so return to (1).

3. There are 15 cards available. Search through all triples and list all *SET*s found. If there is a *SET*, go to (a). If there are no *SET*s, and there are cards left in the deck, add three cards and go to (b). If there are no *SET*s, and there are no cards left in the deck, go to (4).

   a. Choose one *SET* and remove it. Do not add any more cards. Go to (1).

b. There are now 18 cards available. Search through all triples and list all *SET*s found. If there is at least one *SET*, go to (i). If there are no *SET*s, and there are cards left in the deck, add three cards and go to (ii). If there are no *SET*s, and there are no cards left in the deck, go to (4).

    i. Choose one *SET* and remove it. Do not add any more cards. Go to (3).

    ii. There are 21 cards available, so there must be a *SET*. List all *SET*s, choose one, and remove it. Go to (b).

4. You are done.

10.1.1 Why Simulations Are Necessary

Why can't we have the computer run through every possible permutation of the deck, then, for each permutation, run through all possible plays of the game? Here's why: First, there are 81! permutations of the deck. This is approximately 5.8×10^{120}, much larger than the estimate for the number of atoms in the universe.[2]

This has overcounted the number of decks from the point of view of the game. To find the number of different SET decks, first choose 12 cards for the initial layout, then repeatedly choose three cards from the remainder of the deck. This still leaves around 1.5×10^{94} decks, which is larger than any human can comprehend.[3] You can do this calculation for yourself in exercise 10.2.

But we don't need to run through the entire deck to answer some of the questions we care about. For instance, how many initial configurations of 12 cards contain no *SET*s? To answer this question completely, why can't we run through all possible configurations of 12 cards, then count how many times there were no *SET*s? Even this problem is too big: there are $\binom{81}{12} \approx 7.07 \times 10^{13}$ subsets of 12 cards taken from the

---

[2] We are always happy when we see the phrase "number of atoms in the universe." We're not sure why. Physicists estimate the number of atoms in the observable universe to be between 10^{78} and 10^{82}. That means if each atom looked at one deck every millisecond, it would still take more than a quadrillion (10^{15}) years for all those atoms to check all possible decks.

[3] This is a direct challenge to you. Go ahead and comprehend.

entire deck. Running through all those possibilities to look for *SET*s would still take too much time.[4]

Finally, the number of ways to continue play is also too large to be able to enumerate. (Mathematicians use the term *game tree* to describe all possible ways the game can be played.) For instance, for a single ordering of the deck, suppose there are two *SET*s available in each 12-card layout you encounter in a game (no matter which *SET*s were taken during the game). If the game has 24 rounds, we would have $2^{24} = 16,777,216$ possible plays of the game for one single such deck.

Takeaway message: We need simulations.

10.2 NUMBER OF *SET*S IN A LAYOUT

As you play SET, the number of *SET*s on the table at any given point has a huge influence on how quickly the game proceeds. In this section, we'll find the frequencies for the differing numbers of *SET*s in the initial layout, then investigate how things change over the course of a "typical" game.

10.2.1 *SET*s in the Initial Layout

To get data, we simulate the beginning of the game 100,000,000 times. Our simulation first "shuffles" the deck, and then selects 12 cards and counts the number of *SET*s present. (*SET*s are allowed to overlap here; if the same card is in two or more different *SET*s, each of those *SET*s is counted.) The number of *SET*s in those initial layouts varies from 0 (the minimum) to 14 (the maximum). Table 10.1 gives the results of the trials; the same information is shown graphically in figure 10.1.

How can we tell if the simulation is giving reasonable answers? There are a few ways to check. We begin by verifying something we already know: the expected number of *SET*s in the initial layout of 12 cards.

---

[4] Doing 100,000,000 simulations of the game on a MacBook Pro takes more than an hour. This problem is about 70,000 times larger—at this rate, it would take around eight years to exhaust all the cases. But this might be feasible with streamlined code and faster machines or parallel processing.

TABLE 10.1.
The number of *SET*s in the first layout (100,000,000 trials, random *SET* removal).

# SETs	# Layouts	Percentage
0	3,228,460	3.2%
1	14,519,427	14.5%
2	26,096,625	26%
3	27,258,094	27%
4	18,024,022	18%
5	7,989,819	8%
6	2,331,884	2.3%
7	468,357	< 0.5%
8	68,288	≈ 0.07%
9	11,659	≈ 0.01%
10	2964	very small
11	229	very, very small
12	137	tiny
13	31	very tiny
14	4	very, very tiny

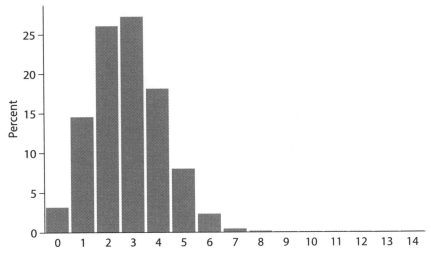

Figure 10.1. Percentage of times there was a given number of *SET*s in the first layout of 12 cards.

If we calculate the average number of *SET*s from the trials in our simulation, our answer should be very close to $220/79 \approx 2.7848$, the value we calculated in chapter 3 for the expected number of *SET*s in the initial layout of cards.

Using the data from the simulation, we find the average number of SETs is 2.78487..., very close indeed to the predicted value of 2.78481. This gives us some confirmation that our simulation is doing what we thought it was doing.[5] Moreover, we can use the standard deviation of the data (around 1.38) to estimate how close our numbers should be to the theoretical values.[6] We find that our approximation should agree in the first 3 or 4 places after the decimal, which it does.

But we have much more information than that. For instance, we can address a question we asked way back in chapter 1.

- How often does it happen that there are no SETs in the opening layout of 12 cards?

This question is also important in playing the game; in fact, the instructions that come with SET say that the odds that there are no SETs in the initial 12-card layout is approximately 33:1, giving a probability of $1/34 \approx 2.94\%$. This claim has received a fair amount of attention on the web, where people have tried to figure out how that number was calculated or run simulations to try to verify it.

In the simulation, we find that the initial layout contains no SETs around 3.2% of the time, slightly more often than the 2.94% claim from the instructions. Our estimate agrees with what other people who have run simulations have posted online.[7]

It's interesting to note that we expect to have either two or three SETs present in the initial layout a little more than 50% of the time. And, while it's possible for more than six SETs to be present in the initial layout, this happens very rarely.

There is another way to verify the accuracy of the simulation: the expected number of times we get 14 SETs in the initial 12-card configuration. (This is the maximum possible. See project 5.1.) We can calculate the expected number of times this will happen in 100,000,000 trials exactly. This theoretical expected value turns out to be about 4.3; since we got 14 SETs 4 times in our 100,000,000 trials, we have another

---

[5] That's a good thing.

[6] This sweeps lots of assumptions under the rug. Watch your step.

[7] That's also a good thing.

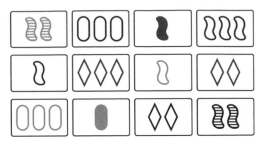

Figure 10.2. A typical SET puzzle. Find the six *SET*s.

reason to believe the simulations.[8] See exercise 10.1 for the details of this calculation.

Finally, we consider one last number from table 10.1: the percentage of times there are exactly six *SET*s among the first 12 cards. This happened about 2.3% of the time in our simulation. But such configurations should be familiar to many readers: these are precisely the layouts you see in the SET Daily Puzzle, with a new puzzle every day available at http://www.setgame.com/set/puzzle (see figure 10.2).

This suggests the following strategy for designing the Daily Puzzle (consisting of 12 cards that contain exactly six *SET*s): Randomly generate a large number of 12-card configurations, say 1000, and count the number of *SET*s in each of them. Then, with probability greater than 99%, you will have at least one configuration that contains exactly six *SET*s, suitable for the Daily Puzzle. (Computing this probability is a standard exercise in using the normal curve to approximate a binomial distribution—see the discussion in chapter 7.)

10.2.2 Counting *SET*s Later in the Game

As we play the game, the expected number of *SET*s in successive layouts changes. Table 10.2 gives the results of 100,000,000 simulated games; figure 10.3 shows the data for the first 24 layouts graphically. For each game, we keep track of the average number of *SET*s in each 12-card layout encountered during the game. In the table, layout 1 corresponds to the initial layout of 12 cards, layout 2 corresponds to the second

[8] If this ever happens to you in a game, you can confidently claim that someone arranged the deck rather carefully before dealing. We'll take your side in the ensuing argument.

TABLE 10.2.
Average number of *SET*s in each 12-card layout (except for the last three layouts):
100,000,000 trials, with *SET*s removed randomly.

Layout	# SETs	Layout	# SETs	Layout	# SETs
1	2.7849	9	2.3755	17	2.3580
2	2.5322	10	2.3724	18	2.3572
3	2.4364	11	2.3696	19	2.3566
4	2.4012	12	2.3668	20	2.3563
5	2.3892	13	2.3642	21	2.3560
6	2.3846	14	2.3621	22	2.3555
7	2.3814	15	2.3606	23	2.3553
8	2.3785	16	2.3593	24	2.3542
25	0.6689	26	0.0509	27	1

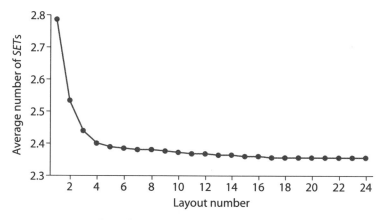

Figure 10.3. Mean number of *SET*s in each of the first 24 layouts (100,000,000 trials, *SET*s chosen randomly).

layout (after a *SET* chosen randomly has been removed and those three cards have been replaced), and so on.

We have a few observations about the average number of *SET*s as the game progresses:

1. The average number of *SET*s in successive layouts drops. This result is perhaps not very surprising to experienced players. As the game progresses, it can become harder to find *SET*s. In fact, it is often the case that you need to add three cards to a layout at some point during the game.

2. We were surprised to see that the average number of *SET*s remains above 2 throughout the game (until we encounter layouts with fewer than 12 cards). This helps explain why the game is so well designed: on average, there will be more than two *SET*s to find for each layout.

3. The largest drop in the average number of *SET*s seems to occur in moving from the first layout to the second (from 2.78 to 2.53, a decrease of roughly 9%). Why should this be true? Here's a partial justification: Note that once a *SET* has been removed from the initial layout, the 117 *SET*s that meet that *SET* are also eliminated from the deck. This reduces the total number of *SET*s in the rest of the deck by more than 10%.

4. The data for the last three layouts are included, but they have been separated at the bottom of table 10.2. For layouts 25 and 26, there are fewer than 12 cards on the table. For layout 27 (if there is one), the three cards must form a *SET*.

Technical notes on the simulation. In running this simulation, we require all layouts (except the ones near the end of the game) to have exactly 12 cards. Of course, it is possible that there are no *SET*s in some 12-card layouts. Here's a short description of how our algorithm gets from one layout to the next. Suppose we have taken $n - 1$ *SET*s, and we now have 12 cards forming layout n. Then layout $n + 1$ is created in one of two ways:

1. There's a *SET*: Layout n has 12 cards, and a *SET* is found. Record the number of *SET*s in the layout. Then remove a *SET* and add 3 more cards. Those 12 cards form layout $n + 1$.

2. There's no *SET*: Layout n has 12 cards, but there are no *SET*s. Record 0 as the number of *SET*s in layout n. Add 3 more cards, then find a *SET* and remove it. This leaves us with 12 cards to form layout $n + 1$.

How could this procedure fail? In the second situation above, it's possible that there are still no *SET*s among the 15 cards. Then we add 3 more cards, bringing the total to 18. Now, assuming there is a *SET* among these 18 cards, we remove a *SET*, reducing the total to 15.

TABLE 10.3.
The number of cards left at the end of the game in 100,000,000 trials.

Cards left	Average number of SETs remaining in the deck	Expected number of SETs in a random collection
18	9.98	10.33
15	5.38	5.76
12	2.35	2.78
9	0.67	1.06
6	0.05	0.25

Assuming there is a *SET* among the 15 remaining cards, we remove that *SET*, bringing the total number of cards back to 12.

In this situation, we have skipped layout $n + 1$ entirely; from the viewpoint of our simulation, we move directly from the 12-card layout n to the 12-card layout $n + 2$. Thus, a layout is not counted in our data any time we need more than 15 cards to continue the game.

How often did this happen? In layout 2, a total of 36,294 of the 100,000,000 trials were not counted. This represents 0.036% of the data, small enough for us to ignore any ambiguity that might arise in discarding these cases.

10.2.3 SETs at the End

The simulation also keeps track of how many *SET*s are left in the remainder of the deck (on the table and not yet dealt) at each layout. For the cards left near the end of the game, this generates some interesting data. Table 10.3 gives the number of cards remaining, the average number of *SET*s those cards contained in the simulation, and the theoretical expected value for the number of *SET*s in that number of randomly chosen cards.

You can see that the cards remaining (on the table and in the deck) contain fewer *SET*s, on average, than a random collection of the same number of cards. We conclude that these collections of cards are not random. Since the cards remaining to be dealt are a completely random collection, the problem must be the cards on the table. Those cards have had opportunities to be removed at different points during the game, but they weren't. The difference between the expected number of

TABLE 10.4.
The number of cards left on the table at the end of the game in 100,000,000 trials.

Cards left	Percentage of outcomes
0	1.22%
3	0%
6	46.8%
9	44.5%
12	7.37%
15	0.077%
18	$5.4 \times 10^{-5}\%$

*SET*s and the actual number of *SET*s is more pronounced when there are fewer cards left, because the cards on the table represent a larger proportion of the total.

10.2.4 Number of Cards Left at the End

We know a game cannot end with 3 cards left on the table. We also know (from the study of maximal caps in chapter 9) that any collection of 21 cards must contain at least one *SET*. So the number of cards remaining on the table at the end of the game could be 0, 6, 9, 12, 15, or 18. How likely are each of these outcomes? Using the same simulation as above, we can answer these questions. Table 10.4 summarizes the results from 100,000,000 simulated games.

As expected, we never find 3 cards left at the end. Also, note that a bit over one time per hundred, no cards are left—we cleared the deck. This is consistent with our experience playing the game. It happens quite infrequently, although we haven't kept track of how often we've cleared the deck over our years and years of play. This gives you another reason to play tons of games.

Finally, note that 12 or more cards remain at the end of a game infrequently (less than 7.5% of the time). This is consistent with our earlier simulation; the final 12 cards on the table contained more than two *SET*s, on average (of course, two *SET*s could share a card). On the other hand, comparing the probability that there are no *SET*s in the initial layout of 12 cards (around 3.2%) with the probability that there will be no *SET*s in the final 12 cards (around 7.4%), we note the latter situation occurs more than twice as often.

10.3 HOW TO REMOVE *SETS*?

In all the simulations that we did in the previous section, the computer was programmed to remove *SET*s by identifying all *SET*s on the table and choosing one at random. This does not necessarily reflect the way that most people play. It is our experience that people usually see *SET*s with the most attributes in common first.[9]

Does this matter? More precisely, we ask the following question.

Question: Does the procedure for removing *SET*s change the expected number of *SET*s in each layout during the game?

Before we ran our simulations, we didn't think so. We ran four different simulations, each one removing *SET*s in one of four different ways:

- **Random:** Choose a *SET* at random (as in the last section).
- **MostAttributes:** Select the *SET* with the most attributes the same (and choose randomly in the case of ties).
- **Lexicographic:** List all cards as 4-tuples and order them using lexicographic order.[10] Take the *SET* with the earliest card in lexicographic order (and use the second card in the *SET* if that card is in more than one *SET*).
- **SetSum:** First, convert all the cards to ordered 4-tuples, then add up the four coordinates (not mod 3—just compute the sum) for each card. Call this the *CardSum*. The *SetSum* is the sum of the CardSums for the cards in the *SET*. For example, if the three vectors are (1,2,1,0), (2,1,1,2), and (0,0,1,1), then the three

---

[9] We have only anecdotal evidence for this. One of the authors claims that she usually sees the all-different *SET*s first. The rest of us aren't sure what to think, but she is fast enough at the game that we believe her.

[10] *Lexicographic* order is sometimes called "alphabetical" order: the earliest card is the one with the lowest number in the first coordinate. In the case of a tie, take the one with the lowest number in the second coordinate, and so on.

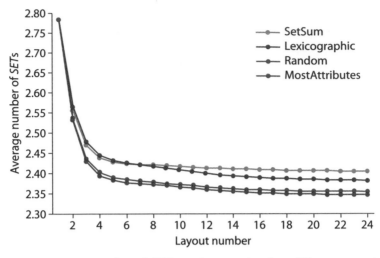

Figure 10.4. Average number of *SET*s per layout using four different strategies in 100,000,000 trials.

CardSums are 4, 6, and 2, so the SetSum is 12. Remove the *SET* with the smallest SetSum. (In the case of a tie, choose randomly.)

Will all four ways of removing *SET*s give approximately the same average number of *SET*s in each layout? If not, which procedure will give the largest expected values and which one will give the smallest? (It's worth thinking about this before plunging heedlessly ahead.) The results from our simulations are shown in the graph in figure 10.4.

We were surprised by the findings.[11] Different strategies for choosing *SET*s give different expected numbers of *SET*s as you play the game. Each starts at the same point (2.78 *SET*s in the first 12-card layout), but even after taking the first *SET*, we see a difference in the expected number of *SET*s in the second layout. Why should it matter which algorithm you use to remove the first *SET*? We find this puzzling.

Here are a few observations about the four simulations based on the data in figure 10.4:

1. Removing *SET*s using the SetSum strategy seems to be the best for having the most *SET*s available as you play the game, while the MostAttributes algorithm seems to leave the fewest.

[11] "Astonished" might be a better description of our reaction. Programmer Brian Lynch was particularly surprised. He'd devised SetSum because it was easy to program, and he genuinely didn't think it mattered.

2. We initially thought the SetSum, Lexicographic, and Random algorithms would produce similar results, but this seems to be false. Removing *SET*s randomly reduces the number of *SET*s available, compared to the SetSum and Lexicographic strategies.

3. As we mentioned earlier, we believe most people play SET by removing *SET*s with more attributes in common. As the graph shows, this way of playing results in fewer *SET*s on the table as the game progresses (compared with the other three strategies explored). This is unfortunate, and it bears further exploration. Is this a consequence of the fact that there are fewer *SET*s with three attributes the same (these represent 10% of all the *SET*s) than there are for the *SET*s with fewer attributes the same?

4. How does changing our coordinate assignment of cards to vectors affect these procedures? On the one hand, the specific coordinates chosen for the cards shouldn't change the overall behavior of any of these procedures, i.e., we would expect our graph in figure 10.4 to be the same. In fact, the coordinates chosen will have no effect on the Random and MostAttributes procedures. But changing the coordinate system will radically change how *SET*s are selected using the SetSum or Lexicographic methods.

It might be interesting to stop each simulation after 20 rounds, say, to measure how "skewed" the cards on the table are. One way this might be measured is to list all the vectors for all the cards in the layout, then take the mean of each coordinate. We would expect the different procedures to yield different results, on average.

Results similar to ours can be found on H. Warne's blog (http://henrikwarne.com/2011/09/30/set-probabilities-revisited/, a response to a blog post by P. Norvig at http://norvig.com/SET.html). Warne's simulation compared removing the first *SET* found (likely to be very close to our Lexicographic procedure), finding all *SET*s and taking one randomly (our Random procedure), and removing the *SET* with the most attributes the same (our MostAttributes procedure). (He did not include the SetSum choice.) As we have, he found a difference between taking the first *SET* found and the other two choices.

On the other hand, we don't need to get too worked up about all of this. The difference between the "best" and "worst" procedures is approximately 0.05, or about a twentieth of a *SET*, and this difference persists throughout the latter stages of the game. Thus, we would expect to see about one more *SET* every 20 games or so when we use SetSum, compared with MostAttributes.

But there is a difference! We conclude this section with a philosophical comment. We know all *SET*s are the same (we showed they were all *affinely equivalent* in chapter 8). But by removing the *SET*s in different ways, the game proceeds differently. Does this mean *SET*s really are different? We encourage you to embrace this mystery, and perhaps attempt to resolve it.

10.4 REMOVING DISJOINT *SETS* FROM THE ENTIRE DECK

One point of this chapter is that you can simulate the game in a variety of ways. But, until now, we've tried to have the computer imitate the way an actual game proceeds, laying out 12 cards, taking a *SET*, replacing the cards taken, and so on.[12] But there are other ways we could have programmed the computer to remove *SET*s from the deck. In particular, what if we simply lay out the entire deck, then take *SET*s at random:

> **Question:** Is there a difference between removing *SET*s from the entire deck or removing *SET*s randomly from 12-card layouts in a series of rounds?

To answer this question, the computer removed *SET*s, chosen at random from the full deck, one at a time, and then counted how many *SET*s remained in the deck at each stage in the process. The results are shown in table 10.5; a graph of the average number of *SET*s remaining per layout is shown in figure 10.5. Different trials of the simulation gave a wide variation in the number of *SET*s left after a given number of *SET*s had been removed. In our simulation, we found the largest range occurred after 13 *SET*s had been removed, with a maximum of 168 *SET*s remaining on the board and a minimum of 106 *SET*s remaining.

---

[12] But much faster—it's a computer.

TABLE 10.5.
100,000,000 trials, *SET*s chosen randomly: average number of *SET*s left in the deck
after disjoint lines have been removed.

SETs removed	Average number of SETs left	Maximum	Minimum	SETs removed	Average number of SETs left	Maximum	Minimum
0	1080	1080	1080	14	113.86	142	83
1	962	962	962	15	88.75	115	61
2	853	853	853	16	67.62	94	43
3	752.63	753	744	17	50.10	94	28
4	660.52	662	653	18	35.87	61	19
5	576.31	580	562	19	24.57	48	8
6	499.62	507	479	20	15.87	34	4
7	430.11	443	408	21	9.45	26	1
8	367.39	384	333	22	4.96	22	1
9	311.11	330	276	23	2.08	13	1
10	260.92	281	225	24	0.59	6	1
11	216.45	239	180	25	0.04	2	2
12	177.36	201	141	26	1	1	1
13	143.28	168	106				

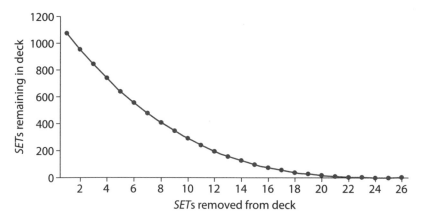

Figure 10.5. Average number of *SET*s remaining in the entire deck when *n* *SET*s have
been removed.

The difference between playing the game and simply removing *SET*s
from the whole deck has been discussed on the web. For example, on
StackExchange's MathOverflow website, there was a discussion of the

TABLE 10.6.

Comparing the frequencies for different numbers of cards left at the end of the game using two different simulations of 100,000,000 trials.

Cards left	Percentage of outcomes, playing the game	Percentage of outcomes, removing SETs from whole deck
0	1.22%	0.878%
3	0%	0%
6	46.8%	41.46%
9	44.5%	47.25%
12	7.37%	10.26%
15	0.077%	0.151%
18	5.4×10^{-5}%	1.68×10^{-4}%

odds of having a "perfect" game of SET, i.e., a game where the entire deck is cleared. Here is a question from that discussion:

> Say all 81 cards are laid down face up, and *SET*s are removed randomly until no *SET*s remain. Is the probability of having no left over cards the same as it is when the game is played normally? (anonymous questioner, MathOverflow at StackExchange, http://mathoverflow.net/questions/66400/probability -of-having-a-perfect-game-of-set)

In the discussion, Warne said he leaned toward there not being a difference, but he wasn't completely convinced. Our data indicate that the probabilities are different.

In table 10.6, we compare the data obtained from the simulation in section 10.2 (when the computer plays the game as usual) with the data from the simulation above (when *SET*s are chosen randomly from the entire deck).

We conclude that there is a difference between these two procedures—taking *SET*s randomly from the entire deck increases the number of cards remaining at the end of the game. The average number of cards left at the end of the game for the simulation from section 10.2 is around 7.7; for the simulation in this section that removes *SET*s from the entire deck, it's around 8.

Moreover, the chances of clearing the deck are dramatically differ-ent.[13] Although the absolute difference between these two probabilities

[13] Maybe we're being melodramatic here, but your authors are an expressive group.

is 1.22%–0.878% = 0.342%, which is small, the right way to look at this is through ratios:

$$\frac{1.22\%}{0.878\%} \approx 1.4.$$

Thus, we are approximately 40% more likely to clear the deck when playing the game, compared to removing *SET*s randomly from the entire deck. This addresses the question from the MathOverflow website.

We conclude this section with a final puzzler. Here are two situations to compare.

1. Choose a *SET* and put it aside. Now, deal 9 cards from the rest of the deck. How many *SET*s are there, on average, among those 9 cards? This can be calculated exactly: the theoretical expected value is 1.0622.

2. Next, deal 12 cards from the deck, and remove one *SET* at random from those 12. (If there are no *SET*s present, pick up the cards, shuffle, and redeal.) What is the expected number of *SET*s in the 9 cards that remain? This cannot be calculated easily, but in our simulation, we found an average of 0.8147 *SET*s among those 9 cards. (In the simulation, collections of 12 cards that had no *SET*s were discarded.)

This difference is significant, both practically and statistically. We see that there is a fundamental difference between first dealing, then removing a *SET*, versus first removing a *SET*, then dealing. This is worth further exploration; see projects 10.2 and 10.3.

10.5 THE LAST SIX CARDS

Suppose you have played a game of SET, and there are six cards left. These cards will have an interesting property: when you randomly partition them into three pairs, either the three cards that complete *SET*s with each pair must themselves form a *SET*, or the same card completes each of these three *SET*s. (We've addressed this situation twice—see exercise 1.2 or the extended discussion in section 5.7.) We also note that the six cards that remain must sum to $\vec{0}$.

Our simulations allow us to address the next question:

Question: Among all configurations of six cards that sum to $\vec{0}$ and that do not include any *SET*s, what proportion can be partitioned into three pairs so that the same card completes the *SET* for each pair?

If we can partition the six cards into three pairs so that the same card completes the *SET* for each pair, we have a triple interset. If we assume each legal configuration of six cards (meaning those cards sum to $\vec{0}$) is equally likely to occur, we can calculate (no simulations needed!) how often this happens: about 21.74% of all the legal configurations of six cards are a triple interset.

But is each legal six-card configuration equally likely to appear? To check this, we can run another simulation. This simulation has two parts.

1. The first part is what we've already done several times, with a twist: just play the game, repeatedly removing random *SET*s, but keep only those trials that result in six cards at the end.
2. Check if those six cards form a triple interset.

When this simulation was run with 100,000,000 trials, with *SET*s removed randomly, the six cards at the end of the game formed a triple interset 18.1545% of the time. This means that the assumption that these configurations occur with equal probabilities is wrong, at least if the *SET*s are removed randomly.

What can we conclude? Since the triple interset possibility appears less frequently than we'd expect, that kind of configuration must be destroyed more often in playing the game. To explore further, we can repeat the simulation, but use different ways to remove *SET*s—see exercise 10.8.

10.6 THE END GAME

We play the End Game every single time[14] we play SET. At the beginning of the game, one card is set aside face down. You then play

---

[14] Here's why: We love the End Game.

TABLE 10.7.
Percentage of times the game ends with *n* cards, and the probability of the End Game
being won in each case. Note that 17 cards were left in 309 of the 100,000,000 trials.

# cards remaining	Percentage	Win percentage
2	1.34%	100%
5	25.7%	0%
8	55.9%	39.54%
11	16.6%	59%
14	0.52%	85.4%
17	(309)	86%

the usual way to the end, until there are no *SET*s left to be found among the cards remaining on the table. Then you figure out the card that you set aside at the beginning of the game from the cards that are left on the table. If you are the first to find a *SET* containing that card, you "win" the End Game. Unfortunately, it doesn't always happen that the card makes a *SET*, so in that case, no one wins.

- How often is there an End Game winner?

Unlike the situation in the last section, even if we assume that all final layouts are equally likely, calculating the probability that the End Game produces a card that completes a *SET* with the cards left on the table would be extremely difficult, if not impossible. So this situation is eagerly waiting for a simulation.

For this simulation, we remove one card from the deck, then play the game until we reach the End Game. At the end, we find out if that card forms a *SET* with two of the remaining cards. Table 10.7 shows the results of these simulations.

As we've seen before, when two cards are left, there must be a *SET* with the missing card, and when five are left, there can't be a *SET*. When there are more cards, your chances of finding a *SET* with the missing card increase. Of course, the more cards there are, the harder it is to figure out the missing card. Presumably, this makes playing the End Game more rewarding—you're more likely to find a *SET* under those conditions.

We can use the data in table 10.7 to answer our motivating question:

- The overall probability of the End Game being won is about 33.6%.

There's another intriguing bit of data in the table. Note that the situations when the entire deck is cleared correspond precisely to the times there are two cards left. This happens 1.34% of the time, according to our simulation. But in our original simulation back in section 10.2, we cleared the deck 1.22% of the time. This means that if you put a card aside, it is more likely that you will be able to clear the deck than if you hadn't removed the card.

- How does putting a card aside help you clear the deck?

We don't know, and it gets worse.[15] We ran this simulation four times; in addition to using our Random procedure, we also ran 100,000,000 trials for each of the MostAttributes, SetSum, and Lexicographic algorithms from section 10.2. The percentages for each of these procedures are very close to the percentages in table 10.7. But the SetSum and Lexicographic procedures each cleared the entire deck 1.47% of the time, higher than the 1.34% above, and a 20% increase from the 1.22% we found in our original simulation (which did not put a card aside).

10.7 CAN YOU ALWAYS CLEAR THE DECK?

There once was a website that had a (non-approved) version of SET. Every day, a new deck was generated, so everyone played the same deck. The game proceeded as a solitaire version of SET—12 cards were dealt, a *SET* was taken, new cards were dealt, and the game proceeded until there were no more *SET*s to take.

The website also included lots of data: For each person who played, they listed the username, how long they took to play the deck, and how many *SET*s they took. In addition, the website kept track of the top ten players (best average times over the course of a month) and the top ten

---

[15] Or better, depending on whether you want more puzzles.

times overall. Often the monthly top ten times ranged from less than a minute and a half to between two and a half to three minutes; the best-ever time was 1 minute, 7 seconds.[16]

The data showed that some decks were clearly harder to play than others: for some decks, the top ten players didn't play it at all, possibly because those players started the game, but quit before the deck cleared.[17]

This leads to a natural question:

- Given one particular deck, how many different ways are there to play the game?

You can imagine that if a layout has two SETs that share a card, then taking one of those SETs could lead to a different game compared to taking the other. How different will those games be? This is a question worth exploring. One way to think about this is to play the same deck over and over. Exercise 10.7 asks you to run a simulation to do this. Even better, you could write a program to keep track of the game tree. If you do so, let us know.

There was another intriguing piece of data from the non-approved site. It happened fairly regularly that one player (or, much less commonly, two) would clear the entire deck. This raises another natural question:

- Given a particular deck, is it always possible that there is some sequence of SETs you could take that will clear that deck?

The answer is no, and we give an example, starting in figure 10.6. In this example, each layout of 12 cards has exactly one SET, leading to a final layout of 12 cards with no SETs. Since there is only one SET to take at each stage of the game, there is no way to clear the deck.

We encourage you to go through the successive layouts, in figures 10.6 to 10.8. Here is a tip to make finding SETs a little easier:

---

[16] That's pretty quick. Since there would be 25 SETs or so taken in a game, and each SET consists of three cards, and each card would need to be selected on the computer, this requires around 75 card choices, which means more than one click per second.

[17] One of us must admit that she did this, more than once.

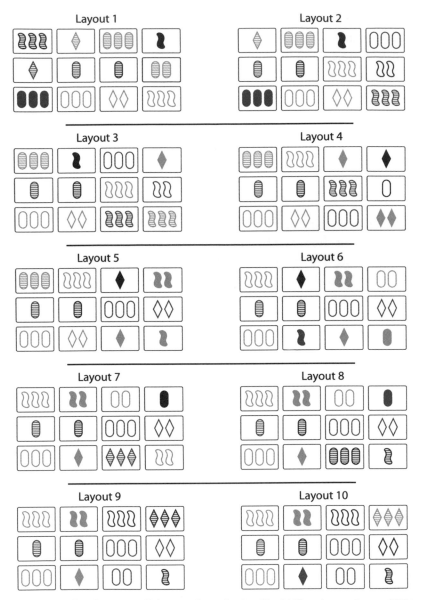

Figure 10.6. Play this game all the way through. Good luck! There's exactly one *SET* in each layout.

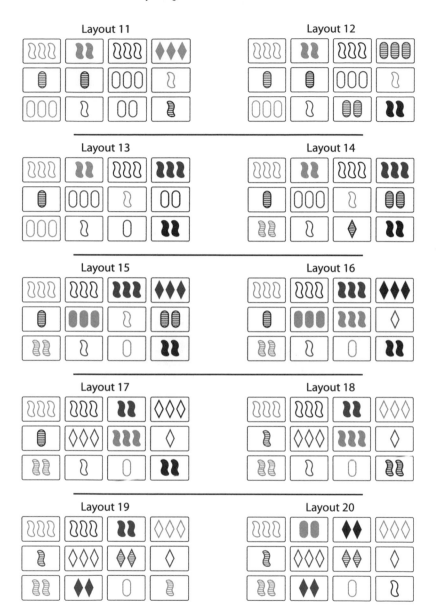

Figure 10.7. Keep playing.

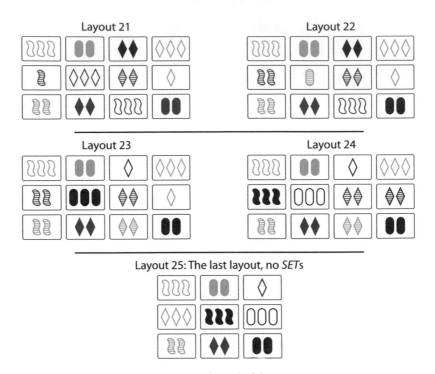

Figure 10.8. The end of the game.

Since you know each layout has only one *SET*, once you find that *SET*, you know that there are no *SET*s in the remaining cards. That means that in the next layout of cards, the (unique) *SET* must contain one of the new cards that are dealt into the positions occupied by the previous *SET*.

As you go through the layouts, you may notice a few things. A few cards stick around for quite a while. In fact, 3 Green Empty Squiggles is in the initial layout of 12 cards, and it survives until the end. More generally, how long does a typical card stick around?

- On average, how long does a given card stay on the table?

This is a question we hope you can explore on your own.[18]

---

[18] Please write to the authors with your results. Some of them will get very excited.

10.8 FINAL WORDS

We hope this chapter has raised more questions than it has answered. It hardly seems necessary to note that computers are an important tool in discovering new mathematics, but we hope that's one of the messages that gets through as you read about the various simulations we've run.

Once we started running simulations, we realized there was more to be done. We are particularly intrigued by the differences we see when *SET*s are removed in different ways; we don't understand yet why this makes a difference, and we want to. Some of our questions are explored in the exercises and projects that follow, but we sincerely hope you've asked other questions we didn't ask.

CALCULATION EXERCISES

The exercises in this chapter are a little different, as all but the first two require computer programming. The exercises generally require only minor changes to the programs that have already been written, while the projects require new ideas and new programs.

EXERCISE 10.1. This problem completes a calculation from section 10.2.1. It is possible for 12 cards to contain 14 *SET*s. (This was explored in project 5.1.) To do this, start with a plane. It is now possible to add three cards to bring the total number of *SET*s to 14.

a. How many ways can you choose 12 cards that contain 14 *SET*s?
b. Compute the probability that 12 randomly chosen cards will have this structure.
c. Compute the expected number of these structures you would encounter in 100,000,000 trials.

EXERCISE 10.2. This problem explores how many different SET games there are. (We used this calculation in section 10.7.)

a. First, how many different ways are there to lay out 12 cards, then 3 cards, then 3 cards, etc.? (This is asking for the number of different decks that are possible from the viewpoint of the game.)

b. For each of the layouts, use the expected number of *SET*s from one of the simulations in this chapter to estimate the total number of games possible for a given deck of cards.

COMPUTER SIMULATION EXERCISES

EXERCISE 10.3. How does the probability that there are a given number of *SET*s in each layout change over the course of a game?

a. First, run a simulation that keeps track of the number of *SET*s in each layout of a game, where *SET*s are removed randomly. How does the number of times there are no *SET*s on the table change over the course of the game? How does the number of times there are 14 *SET*s on the table change over the course of the deck? What happens with the number of *SET*s in between?

b. Now, rerun the simulation where the *SET*s are taken in different ways. Are the results significantly different from when the *SET*s were removed randomly?

EXERCISE 10.4. How much of a difference does the way we remove the *SET*s affect the sizes of the final layouts? This problem asks you to explore this. Rerun the simulations that play the game from section 10.2 in the different ways that were described in section 10.3.

a. Do a simulation that removes the *SET*s chosen randomly.

b. Do a simulation that removes the *SET*s that share most attributes in common (with ties, choose randomly).

c. Do a simulation that removes the *SET*s with the first card in lexicographic order (with ties, choose either randomly or by selecting the *SET* with the second smallest card).

d. Do a simulation that removes the *SET*s with the smallest SetSum (again, choose randomly with ties).

e. Compare the results you got.

EXERCISE 10.5. This problem explores the ways that *SET*s are removed from the entire deck.

a. Run two simulations that remove *SET*s from the entire deck. Stop the simulations after 13 *SET*s (which means that just under half the deck has been removed) and count the total number of *SET*s in the cards that remain. (In (i), you need to run only one simulation, but in (ii) you should run millions.)

i. Run a simulation that removes *SET*s using lexicographic ordering. You need to do only one of these, as the *SET*s will be removed the same way every time. Explain why.
ii. Run simulations that remove *SET*s using the SetSum ordering.

b. Now run two more simulations. Again, stop the simulations after 13 *SET*s (or close to half the deck) and look at the cards that remain. How many of each expression of attributes are there within the cards left?

i. Run simulations that remove *SET*s by the number of attributes in common.
ii. Run simulations that remove *SET*s randomly.

c. Are there significant differences between the answers you get for the two pairs of simulations in parts (a) and (b)? Are there further questions you can ask? [Hint: The answer to the last question is yes.]

EXERCISE 10.6. In section 10.4, we ran a simulation of taking *SET*s from the entire deck, randomly choosing the *SET*s to remove. (Note that if you remove the *SET*s using lexicographic order, then the *SET*s will be removed the same way every time.) Rerun the simulation by choosing *SET*s to remove the other two ways. In each case, have the computer list the *SET*s in the order they are taken.

a. Remove *SET*s by taking those with the most attributes in common (in the case of ties, remove randomly). Keep track of how many *SET*s taken had three attributes the same, two attributes the same, etc.
b. Remove *SET*s by taking those with the lowest SetSum (in the case of ties, choose randomly). What are the SetSums of the *SET*s removed?
c. How much variation is there in the simulation results?

EXERCISE 10.7. Given one particular deck, how many different ways are there to play the game? Explore the answer to this question here.

a. Have the computer choose a random deck. Play the game repeatedly with the same deck. How many *SET*s are left at the end in each trial?

b. Figure out a way to keep track of the actual paths through the game. Warning: This will get huge for your garden-variety deck, so you'll need lots of computer memory and a good way to visualize it.

c. Figure out a way to generate your own deck with a single *SET* per layout.

EXERCISE 10.8. When six cards are left at the end of the game, sometimes you can pair them up so they make a triple interset, and sometimes you can't. In section 10.5, we found that when *SET*s are removed randomly, not all configurations of six cards that contain no *SET*s and sum to $\vec{0}$ are equally likely. Is the same true if you remove *SET*s in different ways?

a. Write a program to play the game, removing *SET*s using lexicographic ordering, but discard any trials that don't end in six cards. Then write a subroutine that tests whether those six cards contain a triple interset. Keep track of the results.

b. Repeat (a), removing *SET*s by the number of attributes in common.

c. Repeat (a), removing *SET*s using SetSum.

d. Discuss your results.

EXERCISE 10.9. Run a simulation to test the average amount of time a card stays on the table during the course of a game. Does it matter which way cards are removed?

EXERCISE 10.10. This question was mentioned in chapter 4. When there are six cards left at the end of the game, it is possible that they all share one attribute. For example, they could all be purple, or they could all be striped. It's our guess that this is pretty rare. Run a simulation to test this. How many configurations ending in six cards share one attribute? How many share two attributes? Do any share three attributes?

PROJECTS

PROJECT 10.1. This project concerns the situation that arises when you are playing a game of SET and there's a layout that has no *SET*s in it. In our house, when we hit upon this situation, we don't add the three cards all at once; rather, we add them one at a time and see if there's a *SET* with the new card. We've found that you typically don't need all three of the cards to find a *SET*. Design and run a simulation to explore the following questions, plus others you might think of.

a. What percentage of games never need three additional cards during the play?

b. What percentage of games require three additional cards more than once?

c. What percentage of games have a situation where the number of cards on the table reaches 18 cards? 21 cards?

d. When games do have a layout that requires additional cards, on average, which layout is the first that did so?

e. If a layout required additional cards, does the next layout require additional cards more frequently, less frequently, or approximately the same?[19]

f. When a layout requires additional cards, what percentage of those times did the first card complete a *SET*? What percentage of those times did the first not complete a *SET* but the second did? What percentage needed a third card? a fourth card? a fifth?...?

PROJECT 10.2. This project explores the difference between (1) taking a *SET* and then dealing and (2) dealing and then taking a *SET*, as discussed at the end of section 10.4.

a. Choose one *SET* (since all *SET*s are the same, you can choose any *SET*) and remove it from the deck, leaving 78 cards.

  i. How many *SET*s do those 78 cards contain? Choose 9 cards from the 78 that remain. Calculate the expected number of *SET*s in those 9 cards, as in chapter 3.

  ii. Run the following simulation: Begin the usual simulation for playing SET. In the first layout, find all the *SET*s. If there are none, skip what follows and redeal. If there is a *SET*, remove one *SET* chosen randomly, but do not replace the cards. Compute the average number of *SET*s contained by the 9 remaining cards.

  iii. Compare the numbers you got in (i) and in (ii). What does the difference tell you?

---

[19] We sometimes find that the layouts following one where additional cards were needed are particularly unpleasant.

b. Now start to play the game, two different ways.

    i. Choose one *SET* (since all *SET*s are the same, you can choose any *SET*) and remove it from the deck. Choose 12 cards from the 78 to start the game (after the first *SET* has been removed). Calculate the expected number of *SET*s in this layout of 12 cards, as in chapter 3.

    ii. Run the following simulation: Begin the usual simulation for playing SET. In the first layout, find all the *SET*s. If there are none, skip what follows and redeal. If there is a *SET*, remove one chosen randomly, replace the cards, and compute the average number of *SET*s in the 12 cards.

    iii. In section 10.2, the simulation we did counted the expected number of *SET*s in two scenarios: (1) when there was a *SET*, we removed it, and replaced it with three new cards, and (2) when there were no *SET*s, we added three cards, and found a *SET* in that collection of 15 cards. In that simulation, we got an expected number of *SET*s in the new layout of around 2.53. Compare three numbers: 2.53, the number from (i), and the number from (ii), which included the number of *SET*s from scenario (1) only. What does the difference tell you?

PROJECT 10.3. This project continues the ideas of the previous project, but in a slightly different way. Here, we ask about the difference between a layout of 12 cards that is guaranteed to contain a *SET* and a layout of cards generated by putting a *SET* on the table and then adding 9 cards from the cards that are left in the deck. Brian Lynch, who ran simulations from project 10.2, came up with two hypotheses that could explain the differences seen in that project. Hypothesis A was that adding a given *SET* to a random collection of 9 cards would yield more *SET*s on average than 12 cards that are guaranteed to contain at least one *SET*. Hypothesis B was that the average number of *SET*s that one card in a (particular) *SET* from a layout of 12 cards containing a *SET* would be greater than the average number of *SET*s that one card in a *SET* chosen from the deck would be in from 9 cards chosen from the rest of the deck.

a. Create two setups for simulations, and run two different simulations for each.

    i. Let setup A consist of one specific *SET* (the same as before, the first *SET* in the ordering) added to 9 cards chosen randomly from the deck.

    ii. Let setup B consist of 12 cards with at least one *SET*.

   iii. Now, run two different simulations:

      I. Have simulation 1 count the average number of *SET*s in each of setup A and setup B.

      II. Have simulation 2 count the average number of *SET*s that each card in the chosen *SET*s of setup A and setup B completes.

b. Use your results to corroborate the results in project 10.2 where results were obtained by counting the average number of *SET*s in 9 cards under two scenarios. The average number of *SET*s using one card from each setup can be tripled to count the average number of *SET*s meeting the three cards in that *SET*. Then the *SET*s remaining are the original *SET* plus all the *SET*s that meet the original *SET*, plus the *SET*s that remain. How close are your results?

   i. Scenario 1 was where one *SET* was removed from the deck and then 9 cards were dealt. This corresponds to setup A, where we chose one *SET* from the deck and then added 9 cards.

   ii. Scenario 2 was where we removed a *SET* from 12 cards that contained a *SET*. This corresponds to setup B, where we looked at 12 cards containing a *SET*.

c. Now you can use a technique similar to the incidence counts from chapters 3 and 7 to count as many of these as you can.

d. Finally, analyze your results. What do you think the difference between the two situations really boils down to?

Conclusion

This book has been a labor of love for the authors. We hope that you have a new appreciation for the game, for mathematics, and especially for the connections between them. As we said early in this book, we have found that mathematics has enhanced our appreciation of the game and that the game has enhanced our appreciation of the related mathematics.

When the game was introduced in the early 1990s, many mathematicians realized that SET was a model for the finite geometry AG(4,3). Because finite geometry is an extremely well-studied area of mathematics, some researchers thought that there was nothing new to be learned from the game. This idea misses several important points, however: Finite geometry treats all *SET*s as the same, but we have seen that this is not true, both in the play of the game, and in the related mathematics. Furthermore, the visualization of AG(4,3) provided by the game has led to new results in geometry, results that almost certainly would not have been discovered without SET. Finally, the game provides an inviting way to get more people interested in geometry, and, more broadly, mathematics. That is a major theme of this book.

And of course, there is so much more that can be done. SET has been a source of family fun, but also research projects. We find that we play differently now: we are often asking questions while we play, so that the game is more of a puzzle than a competition. We hope that as you read, you begin to come up with your own questions to investigate. This book is certainly not the last word on SET.

Remember that all of this came from a simple card game. We encourage you to find new questions to answer and to let us know what you discover. Does fame await? Fortune? Probably not, but there's plenty of fun to be had.

⬦⬦⬦⬦⬦⬦⬦⬦⬦⬦⬦⬦⬦⬦⬦⬦⬦⬦⬦⬦⬦⬦⬦⬦⬦⬦⬦⬦

Solutions to Exercises

CHAPTER 1

1.1. 1GOD

1.2. It works!

1.3. The answer is in figure S.1.

1.4. Count the *SET*s in the 12 cards in figure S.2.

1.5. $27 \times 26/6 = 117$. Choose the first card, then the second card, and the *SET* is determined, but you've overcounted by the number of ways you could choose the same *SET*.

1.6. a. See figure S.3. Then you can find a different ladder.

   b. $3 \times 4 \times 2 = 24$. First, choose the card that stays the same, then choose the attribute you're going to change, then choose the way to change the expression of that attribute for one of the two other cards.

   c. 8. Choose two *SET*s where each card in the first *SET* differs from each card in the second *SET* in every attribute, and each *SET* consists of cards that are different in every attribute. Then, since one card in the ladder must stay the same, it will take four steps to change the first card of the first *SET* to the second card of the second *SET* while leaving the third card the same. But now, it will take another four steps to change the second card of the first *SET* to the second card of the second *SET*. An example is given in figure S.4.

   d. Since you made up the rules, only you can answer this.

1.7. a. It's wrong in number and in shape.

   b. $(2, 0, 0, 1), (0, 2, 1, 1), (2, 1, 2, 0)$.

   c. The sum is $(1, 0, 0, 2)$. The nonzero coordinates are the first and last, which correspond to number and shape, the attributes that were wrong!

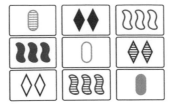

Figure S.1. Exercise 1.3 solution.

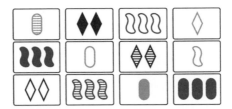

Figure S.2. Exercise 1.4 solution.

Figure S.3. Exercise 1.6(a): One of many solutions.

Figure S.4. Exercise 1.6(c) solution.

d. We think that only an inexperienced player would make a two-attribute mistake. And since the wrong-shaped card is solid, this seems even less likely a mistake.

1.8. a. $A = 2$ Green Empty Ovals, $D = 1$ Green Empty Squiggle. They differ in two attributes.

b. $B = 3$ Red Striped Ovals, $E = 2$ Red Striped Squiggles; they differ in two attributes. The same is true of $C = 2$ Purple Solid Diamonds and $F = 1$ Purple Solid Oval.

c. i. True, because you're correcting the wrong attributes.

ii. False, because the original three weren't a *SET*.

CHAPTER 2

2.1. a. 3 (shading) × 3 (number) × 3 (shape) = 27 red cards.

b. 27 choices for the first card, 26 for the second, with each *SET* overcounted 6 times: $\frac{27 \times 26}{6} = 117$.

c. Given a red card, there are 26 cards it can be paired with. This counts every *SET* twice, so that gives 13 *SET*s.

d. Using the same principle from our original count, we get coordinates using number, shading, and shape: (N, Shd, Shp). We have three attributes and 27 cards that we are choosing from:

- All different: $\frac{27 \times (2 \times 2 \times 2)}{6} = 36$.
- One the same, two different: $\frac{27 \times (1 \times 2 \times 2) \times \binom{3}{1}}{6} = 54$.
- Two the same, one different: $\frac{27 \times (1 \times 1 \times 2) \times \binom{3}{2}}{6} = 27$.

e. Red intersets: $\binom{13}{2} \times 27 = 2106$.

f. Red intersets containing a given card: $\binom{13}{2} \times 27 \times 4 = 27 \times x$, thus $x = 312$.

g. Red planes: As before but with 27 cards, $\frac{27 \times 26 \times 24}{9 \times 8 \times 6} = 39$.

h. Incidence: $39 \times 9 = 27 \times x$, so $x = 13$.

2.2. a. $3 \times 3 \times 3 \times 3 \times 3 = 243$

b. As before: $\frac{243 \times 242}{6} = 9801$.

c. Using the same principle from our original count, we get coordinates using number, color, shading, shape, and feel: (N, C, Shd, Shp, F). We have three attributes and 243 cards that we are choosing from:

- All different: $\frac{243 \times (2 \times 2 \times 2 \times 2 \times 2)}{6} = 1296$.
- One the same, four different: $\frac{243 \times (1 \times 2 \times 2 \times 2 \times 2) \times \binom{5}{1}}{6} = 3240$.
- Two the same, three different: $\frac{243 \times (1 \times 1 \times 2 \times 2 \times 2) \times \binom{5}{2}}{6} = 2430$.
- Three the same, two different: $\frac{243 \times (1 \times 1 \times 1 \times 2 \times 2) \times \binom{5}{3}}{6} = 1620$.
- Four the same, one different: $\frac{243 \times (1 \times 1 \times 1 \times 1 \times 2) \times \binom{5}{4}}{6} = 405$.

d. Given a card, there are 242 cards to pair with it. This counts every *SET* twice, so the answer is 121.

e. Intersets: $\binom{121}{2} \times 243 = 1,764,180$.

f. Planes: As before but with 243 cards, $\frac{243 \times 242 \times 240}{9 \times 8 \times 6} = 32,670$.

2.3. a. $4 \times 4 \times 4 \times 4 = 256$.

b. Take the two cards 1 Empty Red Oval and 2 Striped Green Squiggles. Then we can complete the *SET* in more than one way. Adding 3 Checkered Purple Diamonds and 4 Solid Brown Rectangles makes a *SET*, but so does 3 Solid Purple Rectangles and 4 Checkered Brown Diamonds, for instance.

2.4. Place the four cards in a plane, noting that no three form a *SET*. There are five other cards in this plane, each of which completes at least one *SET* with the given cards. Since there are six pairs of cards, exactly one of those five cards completes *SET*s with two disjoint pairs of the given cards.

2.5. The number of planes containing a given card: $1170 \times 9 = 81 \times x$, thus $x = 130$.

2.6. The number of planes containing a given *SET*: $1170 \times 12 = 1080 \times x$, thus $x = 13$.

CHAPTER 3

3.1. a. There is a 78/79 chance that three cards do not form a *SET*.

b. There are three cards from the remaining 78 that we can't pick, so there are 75 that will work. So the probability that ABD, ACD, and BCD are not *SET*s is $\frac{75}{78}$.

c. The probability the four cards contain no *SET*s is $\frac{78}{79} \times \frac{75}{78} = \frac{75}{79}$.

3.2. Six cards—no *SET*s. Follow the hint:

- Exactly one *SET* is contained: Choose the *SET* in 1080 ways, then choose the other three cards in $\frac{78 \times 75 \times 69}{6}$ ways. This gives $\frac{1080 \times 78 \times 7 \times 69}{6} = 72{,}657{,}000$ ways.
- Exactly two *SET*s are contained:
    - The two *SET*s are disjoint: There are $1080 \times (1080 - 118) = 1{,}038{,}960$ ways to do this.
    - The two *SET*s intersect: Choose the intersection card in 81 ways, then choose the two *SET*s in $\binom{40}{2}$ ways. Finally choose the last card in 72 ways, to give a total of $4{,}548{,}960$.

- Exactly three *SET*s are contained: If the cards are A, B, C, D, E, and F, then we must have *SET*s ABC, CDE, and AEF. Choose the three cards A, C, and E in $\binom{81}{3} - 1080$ ways, then the remaining cards are determined. This gives a total of $84{,}240$.

Adding up all the cases gives 78,329,160 ways to have a bad configuration of six cards. So the probability that there are no *SET*s among the six cards is $(\binom{81}{6} - 78,329,160)/\binom{81}{6} = 75.86\%$.

3.3. a. A triple interset contains an interset. By exercise 2.4, we know that the center is unique.

  b. Choose a card in 81 ways, then choose 3 *SET*s that contain that card in $\binom{40}{3}$ ways. This gives 800,280 triple intersets. To get the probability that 6 randomly chosen cards form a triple interset, divide by $\binom{81}{6}$: this gives $0.00246\ldots \approx 0.25\%$.

  c. Expected number of triple intersets among 12 cards is $(\binom{12}{6} \times 81 \times \binom{40}{3})/\binom{81}{6} = 2.278$.

  d. $EV = 9/79 = 0.1139\ldots.$

3.4. a. There are $\binom{4}{1}2^3$ ways to choose the second card. This gives 32 possible cards, so the probability is $32/80 = 40\%$.

  b. This time, there are $\binom{4}{2}2^2 = 24$ ways to choose the second card, so the probability is $24/80 = 30\%$.

  c. There are $\binom{4}{3}2 = 8$ ways to choose the second card. The probability is 10%.

CHAPTER 4

4.1. These cards don't add up to $(0, 0, 0, 0)$, so they can't be the ones left at the end of a game.

4.2. Stefano is right—everything works fine in the five-attribute game. In particular, two cards determine a unique *SET*.

4.3. a. 1 Green Empty Diamond.

  b. If this formed a *SET* with two of the other cards, there would be three cards left at the end of the game, which is impossible.

4.4. 2 Green Empty Ovals. This forms two different *SET*s with the other cards.

4.5. No. Start with two different cards A and B. If $A + B + C = (0, 0, 0, 0)$, then if $C = A$, for instance, we would have $2A + B = (0, 0, 0, 0)$. But $2A = -A$ (mod 3), so $2A + B = (0, 0, 0, 0)$ is the same as $-A + B = (0, 0, 0, 0)$. So $A = B$, which is a contradiction.

4.6. The center of an interset in unique. Here's why: If ABX and CDX are *SET*s for some card X, then $A + B + X = C + D + X = (0, 0, 0, 0)$. Then

$A + B = 2X$ and $C + D = 2X$, so $A + B + C + D = 2X + 2X = 4X = X$. This means that the center of the interset is the sum of the four cards in the interset, so it's unique.

4.7. If $A + B + C = (0, 0, 0, 0)$, then $C = -A - B = 2A + 2B$ (mod 3). Similarly for the other two equations.

4.8. In all cases, three numbers $\{a, b, c\}$ can occur if they are the same, mod 3, i.e., we can reach $\{0, 0, 0\}$ by repeatedly either subtracting 3 from a, b, or c, or subtracting 1 from each of a, b, and c.

a. When you play the End Game attribute by attribute, there are no other possibilities.

b. $\{9, 0, 0\}$, $\{7, 1, 1\}$, $\{6, 3, 0\}$, $\{5, 2, 2\}$, $\{4, 4, 1\}$, $\{3, 3, 3\}$.

c. $\{12, 0, 0\}$, $\{10, 1, 1\}$, $\{9, 3, 0\}$, $\{8, 2, 2\}$, $\{7, 4, 1\}$, $\{6, 6, 0\}$, $\{6, 3, 3\}$, $\{5, 5, 2\}$, $\{4, 4, 4\}$.

4.9. a. Take the two cards represented by $(1, 1, 1, 1)$ and $(2, 2, 2, 2)$ in the usual assignment. These sum to $(0, 0, 0, 0)$, but they wouldn't if we decided ovals $\leftrightarrow 0$, for instance.

b. Take the four cards $(1, 0, 0, 0)$, $(0, 1, 0, 0)$, $(0, 0, 1, 0)$, $(2, 2, 2, 0)$. These sum to $(0, 0, 0, 0)$, but if our assignment swapped the 0 in the last position with a 1, these four cards would sum to $(0, 0, 0, 1)$.

4.10. a. For example, we could use the pairs (3 Red Checkered Diamonds and 4 Green Checkered Ovals) or (3 Green Checkered Diamonds and 4 Red Checkered Ovals).

b. There are $2^3 = 8$ pairs of cards that complete the "*SET.*"

c. $(0, 0, 0, 0)$, $(0, 0, 2, 2)$, $(2, 2, 0, 0)$, $(2, 2, 2, 2)$.

CHAPTER 5

5.1. Take two intersecting lines L_1 and L_2. Let P be a point not on either line. Show that each line through P either meets both L_1 and L_2, or is parallel to one, but not the other. Show that this gives a one-to-one correspondence between the points on L_1 and L_2. When L_1 and L_2 are parallel, find a third line that meets both L_1 and L_2, then use the first part.

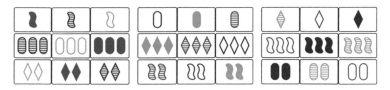

Figure S.5. Exercise 5.5 hyperplane, filled in.

5.2. Get out paper and pencils for this one.

5.3. a. Any card in the right-hand plane completes a *SET* with the 1 Red Solid Squiggle card and a card in the left-hand plane.

 b. We have nine choices for the first card (by part (a)). Then you can check that any ordering of the remaining 16 cards is legal here—each card completes at least one *SET* with the cards already placed.

5.4. a. We know that there are 108 *SET*s that differ in exactly one attribute. By the parallel postulate, there are 27 *SET*s parallel to a given *SET* S (including S). Since $108/27 = 4$, these *SET*s must fall into 4 classes.

 b. This follows from the parallel postulate: Apply the postulate to a given *SET* and each of the 78 cards in the deck that are not in the *SET*. This gives $78/3 = 26$ *SET*s parallel to the given *SET*.

 c. As in (a): There are 324 *SET*s in this category, and $324/27 = 12$ parallel classes.

 d. As in (a): There are 432 *SET*s in this category, and $432/27 = 16$ parallel classes.

 e. As in (a): There are 216 *SET*s in this category, and $216/27 = 8$ parallel classes.

5.5. If the cards are A, B, C, D, E, and F, then assume ABX and CDX are both *SET*s for some other card X. Then let Y be the card that completes the *SET* with D and E. By the work from this chapter, $2X + Y = (0, 0, 0, 0)$, so $X = Y$.

5.6. a. Figure S.5 shows the hyperplane; 3 Red Striped Diamonds is right in the middle.

 b. Let's call the five cards A, B, C, D, and E, where E is the card produced when playing the End Game, i.e.,
$A + B + C + D + E + E = (0, 0, 0, 0)$. Note that
$A + B + C + D = E$. We use the following procedure for creating

new collections of five bad cards, i.e., collections where the End Game fails.

- First, choose three of the first four cards in this list. We choose B, C, and D.
- Then choose a card F to complete a *SET* with cards A and E, so $F = 2A + 2E$.
- Now the five cards B, C, D, E, F are a bad collection. To see this, note that $B + C + D + E + F + F = B + C + D + E + 4A + 4E = A + B + C + D + E + E = (0, 0, 0, 0)$.

We can do this three more times: let $G = 2B + 2E$, completing the *SET* with B and E, let $H = 2C + 2E$, completing the *SET* with C and E, and let $I = 2D + 2E$, completing the *SET* with D and E. Then the same argument shows that each of the following is a bad collection of five cards: $ABCDE$, $BCDEF$, $ACDEG$, $ABDEH$, $ABCEI$.

- In this example, A is 1 Red Solid Squiggle, B is 1 Purple Striped Squiggle, C is 1 Purple Empty Oval, D is 3 Purple Striped Ovals, and E is 3 Red Striped Diamonds. Then F is 2 Red Empty Ovals, G is 2 Green Striped Ovals, H is 2 Green Solid Squiggles, and I is 3 Green Striped Squiggles. Finally, note that the four *SET*s AEF, BEG, CEH, and DEI all have $180°$ symmetry in the hyperplane of figure S.5.

CHAPTER 6

6.1. First choose the k attributes that are the same. For the remaining attributes, how many ways are there to choose the expressions? And how much did we overcount?

6.2. Just do the algebra: $\frac{g(n,k)}{3^{n-1}(3^n-1)/2} = \frac{l(n,k)}{(3^n-1)/2} = \binom{n}{k}\frac{2^{n-k}}{3^n-1}$.

6.3. Choose a card, then choose two *SET*s that contain that card: $3^n(3^n - 1)$ $(3^n - 3)/8$.

6.4. Make the model.

6.5. a. $\frac{g(n+1)}{g(n)} = 9 \times \frac{3^{n+1}-1}{3^{n+1}-3}$. When n is large, the fraction $\frac{(3^{n+1}-1)}{(3^{n+1}-3)} \approx 1$.

b. $\sum_{k=0}^n \binom{n}{k}2^{n-k} = (1 + 2)^n = 3^n$. Remove the $n = k$ term:

$\sum_{k=0}^{n-1} \binom{n}{k} 2^{n-k} = (1+2)^n = 3^n - 1$. Then divide both sides by 2, and multiply both sides by 3^{n-1}.

6.6. a. Use the hint.

b. Again, this is just a lot of algebra.

c. When $n = 1$, we get $\begin{bmatrix} 1 \\ 0 \end{bmatrix}_q = \begin{bmatrix} 1 \\ 1 \end{bmatrix}_q = 1$. Assuming the result for n, we find $q^k \begin{bmatrix} n \\ k \end{bmatrix}_q$ is a polynomial of degree $k(n+1-k)$ and $\begin{bmatrix} n \\ k-1 \end{bmatrix}_q$ is a polynomial of degree $(k-1)(n+1-k)$. Then $q^k \begin{bmatrix} n \\ k \end{bmatrix}_q + \begin{bmatrix} n \\ k-1 \end{bmatrix}_q$ is a polynomial of degree $k(n+1-k)$, as desired.

d. Use the hint:
$$\begin{bmatrix} n+1 \\ k \end{bmatrix}_{q \to 1} = q^k \begin{bmatrix} n \\ k \end{bmatrix}_{q \to 1} + \begin{bmatrix} n \\ k-1 \end{bmatrix}_{q \to 1} = \binom{n}{k} + \binom{n}{k-1} = \binom{n+1}{k}.$$

6.7. a. $3^n(3^n - 1)(3^n - 3)/3!$.

b. $3^n(3^n - 1)(3^n - 3)(3^n - 9)/4!$.

c. $3^n(3^n - 1) \cdots (3^n - 3^{k-2})/k!$.

6.8. a. $81 \times 40 \times 13 \times 4 = 168{,}480$.

b. $3^n(3^n - 1)(3^{n-1} - 1) \cdots (3^2 - 1)(3 - 1)/2^n$.

6.9. a. $\begin{bmatrix} n \\ 2 \end{bmatrix}_3 = (3^n - 1)(3^n - 3)/(3^2 - 1)(3^2 - 3)$.

b. $\begin{bmatrix} n-1 \\ 1 \end{bmatrix}_3 = (3^{n-1} - 1)/2$.

c. $\begin{bmatrix} n-d \\ k-d \end{bmatrix}_3$.

CHAPTER 7

7.1. a. There are $3^n - 1$ potential choices for B, and $3^{n-1} - 1$ of these choices will agree with A in attribute i.

b. $E(X_i) = 0 \times P(X_i = 0) + 1 \times P(X_i = 1) = P(X_i = 1)$.

c. $a_n = \sum_{i=1}^{n} \left(\frac{3^{n-1}-1}{3^n-1} \right) = \frac{n(3^{n-1}-1)}{3^n-1}$.

7.2. a. There are n choices for the champion, and 3^{n-1} ways to distribute certificates.

b. Choose k people in $\binom{n}{k}$ ways, then choose a champion from that group in k ways. Finally, distribute B and C certificates to everyone else in 2^{n-k} ways. Now sum over k.

7.3. a. Rounding, we solve $m^2 - 3m - 3013 = 0$. This gives $m \approx 56$.

   b. The simplified equation is $m^2 - 3m + 2 - 2a3^{n+1} + 12a = 0$. The solution is

$$m = \tfrac{1}{2}\left(\sqrt{8a3^{n+1} - 48a + 1} + 3 \right).$$

7.4. We need at least 44 cards to give an expected number of planes larger than 3.

7.5. a. $q(n, k) = \sum_{i=0}^{k} \binom{n}{i} 2^{n-i}$. Then

$$2q(n, k) + q(n, k-1) = \sum_{i=0}^{k} \left(\binom{n}{i} + \binom{n}{i-1} \right) 2^{n+1-i}$$

$$= \sum_{i=0}^{k} \binom{n+1}{i} 2^{n+1-i}$$

$$= q(n+1, k).$$

   b. Just use the definition of $q(n, k)$ in terms of $p(n, k)$.

   c. Note that $p(n, k+1) > p(n, k)$ from the definition of $p(n, k)$. Then, by part (b),

$$p(n+1, k+1) \approx \tfrac{2}{3}p(n, k+1) + \tfrac{1}{3}p(n, k)$$

$$> \tfrac{2}{3}p(n, k) + \tfrac{1}{3}p(n, k) = p(n, k).$$

   d. Using (b),
$$p(n+1, k) \approx \tfrac{2}{3}p(n, k) + \tfrac{1}{3}p(n, k-1) < \tfrac{2}{3}p(n, k) + \tfrac{1}{3}p(n, k) = p(n, k).$$

7.6. a. Plug $n = 60$ and $p = \tfrac{1}{3}$ into the formulas for the mean and standard deviation.

   b. Using a calculator, the area is approximately 82.9%.

   c. This time, area $\approx$ 86.8%.

7.7. Let $\mu = n(3^{n-1} - 1)/(3^n - 1)$ be the mean and let $g(n, k) = \binom{n}{k} 3^{n-1} 2^{n-k-1}$ be the number of SETs with exactly k attributes the same. Then use the following formula (or something equivalent) for the standard deviation:

$$\sqrt{\frac{\sum_{k=0}^{n-1} g(n, k)(k - \mu)^2}{3^{n-1}(3^n - 1)/2}}.$$

CHAPTER 8

8.1. For example, take the second card: 1 Purple Solid Oval corresponds to $(1, 1, 2, 1)$. Subtracting from $(0, 1, 1, 0)$ gives $\vec{w} = (2, 0, 2, 2)$. Adding $\vec{w}$ to the three vectors in the *SET* gives $(0, 0, 2, 2)$, $(0, 1, 1, 0)$, and $(0, 2, 0, 1)$, as before.

8.2. a. The three vectors in the *SET* are $\vec{u}$, $\vec{v}$, and $2\vec{u} + 2\vec{v}$. Then there are three ways to pair these up to get a direction vector $\vec{d}$:

$$\vec{d}_1 = \vec{v} - \vec{u},$$

$$\vec{d}_2 = (2\vec{v} + 2\vec{u}) - \vec{u} = \vec{u} - \vec{v},$$

$$\vec{d}_3 = (2\vec{u} + 2\vec{v}) - \vec{v} = \vec{v} - \vec{u}.$$

In all three cases, the direction vector is $\pm(\vec{u} - \vec{v})$.

b. Let $S_1 = \{\vec{u}_1, \vec{v}_1, \vec{w}_1\}$ and $S_2 = \{\vec{u}_2, \vec{v}_2, \vec{w}_2\}$ be two *SET*s. If they are parallel, then there is a vector $\vec{z}$ with $\vec{u}_1 + \vec{z} = \vec{u}_2$, $\vec{v}_1 + \vec{z} = \vec{v}_2$, and $\vec{w}_1 + \vec{z} = \vec{w}_2$. Then a direction vector for S_1 is $\vec{d}_1 = \vec{v}_1 - \vec{u}_1$, while a direction vector for S_2 is $\vec{v}_2 - \vec{u}_2 = (\vec{v}_1 + \vec{z}) - (\vec{u}_1 + \vec{z}) = \vec{d}_1$. Conversely, if $\vec{d}_2 = \vec{d}_1$ or $2\vec{d}_1$, then you can check that there is a vector $\vec{z}$ that can be added to each of the three vectors in S_1 to produce the three vectors in S_2.

8.3. This follows immediately from the direction-vector characterization of "parallel" in the solution to the previous exercise.

8.4. Both parts are fun activities. Have fun!

8.5. Behold table S.1.

TABLE S.1.
Exercise 8.5: Table 8.6 completed.

$\vec{x}$	$\vec{y}$	$2\vec{x} + 2\vec{y}$
$\vec{z}$	$2\vec{x} + \vec{y} + \vec{z}$	$\vec{x} + 2\vec{y} + \vec{z}$
$2\vec{x} + 2\vec{z}$	$\vec{x} + \vec{y} + 2\vec{z}$	$2\vec{y} + 2\vec{z}$

8.6. As in the solution to exercise 8.2(b), let $S_1 = \{\vec{u}_1, \vec{v}_1, \vec{w}_1\}$ and $S_2 = \{\vec{u}_2, \vec{v}_2, \vec{w}_2\}$ be two *SET*s. If they are parallel, then there is a vector $\vec{z}$ with

TABLE S.2.
Exercise 8.7

(0,0,0,0)	(0,1,1,1)	(0,2,2,2)
(1,1,2,0)	(1,2,0,1)	(1,0,1,2)
(2,2,1,0)	(2,0,2,1)	(2,1,0,2)

$\vec{u}_1 + \vec{z} = \vec{u}_2$, $\vec{v}_1 + \vec{z} = \vec{v}_2$, and $\vec{w}_1 + \vec{z} = \vec{w}_2$. Then the three cards $\vec{u}_1 + 2\vec{z}$, $\vec{v}_1 + 2\vec{z}$, and $\vec{w}_1 + 2\vec{z}$ complete the plane.

For the converse, if S_1 and S_2 are disjoint *SET*s in a plane, you can check that they must be parallel using the vector representation of a plane in exercise 8.5.

8.7. a. Using vectors, take the two intersecting *SET*s $(0, 0, 0, 0)$, $(0, 1, 1, 1)$, $(0, 2, 2, 2)$ and $(0, 0, 0, 0)$, $(1, 1, 2, 0)$, $(2, 2, 1, 0)$. Then fill in the rest of the plane, as in table S.2.

   b. In the plane, note that the first attribute is the same in each row, the last is the same in each column, the second is the same in the SW–NE diagonals, and the third is the same in the other diagonals.

   c. Each card in the plane has eight cards that differ in one coordinate. If card C is distance 1 from one of these cards, then C is at least distance 2 from any other card in the plane (or else there would be two cards in the plane closer than distance 3). This means $8 \times 9 = 72$ cards are distance 1 from one of the cards here. But that's the rest of the deck.

8.8. a. An interset is determined by two intersecting *SET*s. Since a pair of intersecting *SET*s is sent to another pair of intersecting *SET*s, the interset is preserved.

   b. A plane is determined by two intersecting *SET*s. Now proceed as in part (a).

   c. A hyperplane is determined by three intersecting *SET*s, not all in a plane. These must be sent to three intersecting *SET*s. Since our transformations do not reduce dimension, they must preserve the fact that not all of the *SET*s are in a plane.

   d. This follows from the fact that these transformations are invertible, so if T produced a *SET*, then T^{-1} would take a *SET* to a non-*SET*, which is a contradiction.

8.9. $M = \begin{pmatrix} 0 & 1 & 1 & 0 \\ 0 & 1 & 0 & 2 \\ 2 & 1 & 0 & 0 \\ 0 & 2 & 0 & 2 \end{pmatrix}$ and $\vec{b} = \begin{pmatrix} 1 \\ 1 \\ 0 \\ 0 \end{pmatrix}$.

8.10. Suppose S_1 and S_2 are parallel *SET*s and T is an affine transformation. Let P be the plane containing S_1 and S_2. Then $T(P)$ is a plane (by exercise 8.8(b)), and $T(S_1)$ and $T(S_2)$ don't intersect, so they are still parallel.

CHAPTER 9

9.1. You can't add the three points that complete *SET*s with pairs from the three points, but you can add any of the other three points.

9.2. a. $\binom{4}{2} = 6$; b. $\binom{4}{2}/2 = 3$; c. $81 \times \binom{4}{2} = 486$; d. $81 \times \binom{4}{2}/2 = 243$.

9.3. Either way, send the first anchor to the second, one point from the first cap to a point from the second, and then a point from the other line in the first cap to a point in the other line of the second. This must send the first cap to the second and the unique partition containing the first cap to the second.

9.4. [Hint: You can assume the three planes correspond to the first coordinate. Add the first coordinates of the points in C, and explore the possibilities for nine points.]

9.5. Choose a single maximal cap, as in figure 9.8. Using exercise 9.4, add nine points for a second cap, leaving nine for a third.

9.6. Good luck!

9.7. a. $63 \times 62/6 = 651$; b. $(2^{n+1} - 1) \times (2^{n+1} - 2)/6 = \begin{bmatrix} n+1 \\ 2 \end{bmatrix}_2$.

9.8. a.
  - If the attribute is missing on all cards, all coordinates for that attribute are $(0, 0)$, so they sum to $(0, 0)$.
  - If one expression appears on two cards and the attribute is missing on the third card, then you have $(a, b) + (a, b) + (0, 0) = (0, 0)$ (mod 2).
  - Among the three attributes, one is $(1, 0)$, one is $(0, 1)$, and the third is $(1, 1)$, so their sum is $(0, 0)$ (mod 2).

b. The only way that three numbers can add to 0 (mod 2) is if all are 0 or if two are 1 and the third is 0. Now, concentrate on one attribute: if $(a, b) + (c, d) + (e, f) = (0, 0)$, show that there are only three possibilities and that these correspond exactly to the three conditions.

c. Consider just one attribute. If it's missing on both, it must be missing on the third. If it's missing on one and is an expression on the second, then the third card must have the same expression. If the attribute is the same on both, it must be missing on the third. If the attribute is different on the two cards, the third card must have the third expression.

9.9. We hope you did.

CHAPTER 10: CALCULATION EXERCISES

10.1. a. 3,032,640;   b. 0.000000043;   c. 4.29.

10.2. a. $81!(12!(3!)^{23}) = 1.53 \times 10^{94}$.

b. Multiplying $2.78 \times 2.53 \times 2.4 \times \cdots$ from table 10.2 gives 4.2×10^7.

BIBLIOGRAPHY

We include a comprehensive bibliography so readers can explore further on their own. If you know of something that's missing, please let us know.

WEBSITES THAT HAVE DISCUSSED SET

- www.setgame.com [Of course, our first entry will be the SET website, thanks to Set Enterprises. On the game's website, you can find a Teachers' Corner that includes various papers about SET, and other information about the game.]
- http://www.bluffton.edu/homepages/facstaff/nesterd/java/setendgame.html [Darryl Nester wrote this app to play the End Game online.]
- http://webbox.lafayette.edu/~mcmahone/capbuilder.html [Jordan Awan wrote the Cap Builder, which allows you to create maximal caps, as discussed in chapter 9.]
- http://mathtourist.blogspot.com/2010/07/set-math.html [Ivars Peterson wrote "SET Math," discussing several problems, including maximal caps and affine planes.]
- http://www.ams.org/samplings/feature-column/fc-2015-08 [David Austin wrote a feature column on the American Mathematical Society's web page called "Game. SET. Line," which explored collections of cards with no *SET*s and the probability that the initial configuration of cards has no *SET*.]

WEBSITES WITH VARIANTS OF THE GAME

- http://works.bepress.com/jeffrey_pereira/ [Jeffrey Pereira wrote a paper about a variant he called SuperSET.]
- http://www.zerosumz.com [The website for Zero SumZ, a projective version of SET, created by Alejandro Erickson, Jonathan Lenchner, and Mathieu Guay-Paquet.]
- https://www.ocf.berkeley.edu/~dadams/proset/ [D. Adams also wrote an online version that he calls ProSET.]
- http://socksgame.com/homepage.php [Anna L. Varvak has a version called Socks, which you can play online or purchase.]
- http://stacky.net/wiki/index.php?title=Projective_Set [Anton Geraschenko discusses how he made cards and includes code to make your own.]
- https://jointmathematicsmeetings.org/amsmtgs/2168_abstracts/1106-a1-1254.pdf [Doug Burkholder discussed a different way to make a projective version of SET directly from the SET cards in a talk at the Joint Mathematics Meeting in 2015.]

DISCUSSIONS OF VARIOUS PROBLEMS
IN THE SET UNIVERSE

- http://norvig.com/SET.html [Peter Norvig posted a question on his blog about the odds of finding a *SET* in a layout of 12 cards.]
- http://henrikwarne.com/2011/09/30/set-probabilities-revisited/ [Henrik Warne's blog has a reply entry discussing simulations to find probabilities for various situations in SET.]
- http://mathoverflow.net/questions/66400/probability-of-having-a-perfect-game-of-set [StackExchange's MathOverflow website hosted a discussion of the odds of having a "perfect" game of SET, i.e., a game where the entire deck is cleared.]
- https://terrytao.wordpress.com/2007/02/23/open-question-best-bounds-for-cap-sets/ [Terence Tao's blog discussed the problem of finding the size of maximal caps.]

GENERAL JOURNAL ARTICLES ABOUT SET

- Mike Baker, Jane Beltran, Jason Buell, Brian Conrey, Tom Davis, Brianna Donaldson, Jeanne Detorre-Ozaki, Leila Dibble, Tom Freeman, Robert Hammie, Julie Montgomery, Avery Pickford, and Justine Wong, "Sets, planets, and comets," *College Math. J.* **44**, no. 4 (2013), 258–264. [This article introduces the idea of planets (4 coplanar points) and comets, and then discusses a game based on these definitions.]
- Anna Bickel and Zsuzsanna Szaniszlo, "SET, affine planes and Latin squares," *Math Horizons* (2007) 36–39. [This article introduces the geometry behind the game and connects it to Latin squares, a combinatorial topic.]
- Ben Coleman and Kevin Hartshorn, "Game, set, math," *Math. Mag.* **85** (2012), 83–96. [This article counts the number of ways to place a certain number of cards, up to symmetry, using Pólya theory, a combinatorial topic.]
- Benjamin Davis and Diane Maclagan, "The card game SET," *Math. Intelligencer* **25**, no. 3 (2003), 33–40. [A survey of geometry and other considerations.]
- Norman Do, "Mathellaneous," *Austral. Math. Soc. Gaz.* **31** (2004), 222–233. [This article gives a summary of techniques used to bound the number of points in a cap.]
- Gary Gordon and Elizabeth McMahon, "Error detection and correction using SET," chapter 14 of *The Mathematics of Various Entertaining Subjects: Research in Recreational Math*, J. Beineke and J. Rosenhouse eds., Princeton University Press, 2015. [This covers topics from chapter 8.]
- Hannah Gordon, Rebecca Gordon, and Elizabeth McMahon, "Hands-on SET," *PRIMUS* **23** (June 2013), 646–658. [The article describes some interesting activities that could be used in the classroom, mostly geometric.]

• Judy A. Holdener, "Product-free sets in the card game SET," *PRIMUS* **15** (December, 2005), 289–297. [The author uses a multiplicative structure on the cards as a classroom activity.]

• Anthony Macula, "An analysis of the lines in the three-dimensional affine space over F_3," *Ars Combin.* **52** (1999), 161–171. [This article focuses on counting the number of lines that intersect two complementary piles of cards from the deck.]

• Elizabeth McMahon, "SET in combinatorics/discrete math," to appear in *Recipes for Tactile Learning.* [A classroom activity with SET.]

RESEARCH ARTICLES ON GEOMETRY, MOSTLY ABOUT MAXIMAL CAPS

• Gino Fano, "Sui postulati fondamentali della geometria proiettiva" [On the fundamental postulates of projective geometry], *Giornale di Mat.* **30** (1892), 106–132. [Discusses the projective geometry now called the Fano plane.]

• Raj Chandra Bose, "Mathematical theory of the symmetrical factorial design," *Sankhyā* **8** (1947), 107–166. [First paper to enumerate maximal caps in an affine space.]

• Adriano Barlotti, "Un'estensione del teorema di Segre–Kustaanheimo" [An extension of the Segre–Kustaanheimo theorem], *Boll. Unione Mat. Ital.* **10**, no. 3 (1955), 498–506. [Essentially shows that all caps are affinely equivalent.]

• Gianfranco Panella, "Caratterizzazione delle quadriche di uno spazio (tridimensionale) lineare sopra un corpo finito" [Characterization of the quadratics of a (three-dimensional) linear space over a finite [noncommutative] field], *Boll. Unione Mat. Ital.* **10**, no. 3 (1955), 507–513. [Independently proved the same result as Barlotti.]

• Giuseppe Pellegrino, "Sul massimo ordine delle calotte in $S_{4,3}$" [The maximal order of the spherical cap in $S_{4,3}$], *Matematiche (Catania)* **25** (1970), 149–157. [Finds the size of maximal caps in AG(4, 3).]

• Barbu Kestenband, "Projective geometries that are disjoint unions of caps," *Canad. J. Math.* **XXXII**, no. 6 (1980) 1299–1305. [Finding disjoint unions of caps in projective space, not necessarily maximal.]

• Barbu Kestenband, "Partitions of finite affine geometries into caps," *Linear Multilinear Algebra* **14** (1983) 257–270. [Continues finding disjoint unions of caps, in affine space.]

• Robert Hill, "On Pellegrino's 20-caps in $S_{4,3}$," *Ann. Discrete Math.* **18** (1983) 433–447. [Does the same as Barlotti and Panella, in English.]

• Gary Ebert, "Partitioning projective geometry into caps," *Canad. J. Math.* **XXXVII**, no. 6 (1985), 1163–1175. [Continues looking for disjoint caps in projective space.]

• Yves Edel, Sandy Ferret, Ivan Landjev, and Leo Storme, "The classification of the largest caps in AG(5, 3)," *J. Combin. Theory Ser. A* **99**, no. 1 (2002), 95–110. [Finds the size of maximal caps in AG(5, 3).]

- Jürgen Bierbrauer, "Large caps," *J. Geom.* **76** (2003), 16–51. [A nice survey article about caps.]

- Yves Edel, "Extensions of generalized product caps," *Des. Codes Cryptogr.* **31**, no. 1 (2004), 5–14. [This article gives the lower bound for the size of maximal caps in dimension n.]

- Aaron Potechin, "Maximal caps in AG(6, 3)," *Des. Codes Cryptogr.* **46**, no. 3 (2008), 243–259. [Finds the maximal caps in AG(6, 3).]

- Michael Follett, Kyle Kalail, Elizabeth McMahon, Catherine Pelland, and Robert Won, "Partitions of AG(4, 3) into maximal caps," *Discrete Math.* **337** (2014), 1–8. [Shows that the partitions of the SET deck into maximal caps are not all affinely equivalent.]

- Ernie Croot, Vsevolod Lev, Peter Pach, "Progression-free sets in Z_4^n are exponentially small," available on the ArXiV at https://arxiv.org/abs/1605.01506. [This article introduces a polynomial method in Z_4^n that was adapted by Eilenberg and Gijswijt to give a new upper bound for the cap size.]

- Jordan S. Ellenberg, Dion Gijswijt, "On large subsets of F_q^n with no three-term arithmetic progression," available on the ArXiV at https://arxiv.org/abs/1605.09223. [This brand-new article gives an upper bound for the size of a maximal cap in dimension n.]

- Jordan Awan, Claire Frechette, Elizabeth McMahon, and Yumi Li, "Demicaps in AG(4, 3) and their relation to maximal cap partitions," in preparation. [A substructure of maximal caps provides information about the two affine equivalence classes from the previous paper.]